T0334618

Emerging Technologies in Computing

Emerging Technologies in Computing
Theory, Practice, and Advances

Edited by
Pramod Kumar, Anuradha Tomar, and
R. Sharmila

CRC Press
Taylor & Francis Group
Boca Raton London New York

CRC Press is an imprint of the
Taylor & Francis Group, an **informa** business
A CHAPMAN & HALL BOOK

MATLAB® is a trademark of The MathWorks, Inc. and is used with permission. The MathWorks does not warrant the accuracy of the text or exercises in this book. This book's use or discussion of MATLAB® software or related products does not constitute endorsement or sponsorship by The MathWorks of a particular pedagogical approach or particular use of the MATLAB® software

First edition published 2022
by CRC Press
6000 Broken Sound Parkway NW, Suite 300, Boca Raton, FL 33487-2742

and by CRC Press
2 Park Square, Milton Park, Abingdon, Oxon, OX14 4RN

© 2022 selection and editorial matter, Pramod Kumar, Anuradha Tomar, R. Sharmila; individual chapters, the contributors

CRC Press is an imprint of Taylor & Francis Group, LLC

Reasonable efforts have been made to publish reliable data and information, but the author and publisher cannot assume responsibility for the validity of all materials or the consequences of their use. The authors and publishers have attempted to trace the copyright holders of all material reproduced in this publication and apologize to copyright holders if permission to publish in this form has not been obtained. If any copyright material has not been acknowledged, please write and let us know so we may rectify in any future reprint.

Except as permitted under U.S. Copyright Law, no part of this book may be reprinted, reproduced, transmitted, or utilized in any form by any electronic, mechanical, or other means, now known or hereafter invented, including photocopying, microfilming, and recording, or in any information storage or retrieval system, without written permission from the publishers.

For permission to photocopy or use material electronically from this work, access www.copyright. com or contact the Copyright Clearance Center, Inc. (CCC), 222 Rosewood Drive, Danvers, MA 01923, 978-750-8400. For works that are not available on CCC please contact mpkbookspermissions @tandf.co.uk

Trademark notice: Product or corporate names may be trademarks or registered trademarks and are used only for identification and explanation without intent to infringe.

Library of Congress Cataloging-in-Publication Data
Names: Kumar, Pramod, editor. | Sharmila, R., editor. | Tomar, Anuradha,
editor. Title: Emerging technologies in computing : theory, practice and
advances / edited by Pramod Kumar, Anuradha Tomar, R. Sharmila.
Description: First edition. | Boca Raton : CRC Press, 2021. | Includes
bibliographical references and index. Identifiers: LCCN 2021027695 |
ISBN 9780367633646 (hbk) | ISBN 9780367639501 (pbk) | ISBN 9781003121466 (ebk)
Subjects: LCSH: Computer science—Technological innovations.
Classification: LCC QA76.24 .E44 2021 | DDC 004—dc23
LC record available at https://lccn.loc.gov/2021027695

ISBN: 978-0-367-63364-6 (hbk)
ISBN: 978-0-367-63950-1 (pbk)
ISBN: 978-1-003-12146-6 (ebk)

DOI: 10.1201/9781003121466

Typeset in Minion
by codeMantra

Contents

Editors

Pramod Kumar is Professor, Head & Dean of Research at Krishna Engineering College, India. He previously served as Director of the Tula's Institute, Dehradun, Uttarakhand. With over 20 years of experience in Computer Science, he is a Senior Member of IEEE (SMIEEE), has published research papers in various International Journals and Conferences and authored numerous book chapters, and has organized more than 10 IEEE International Conferences.

Anuradha Tomar is a Postdoctoral Researcher in the EES Group at Eindhoven University, the Netherlands. She is a member of the European Commission's Horizon 2020, UNITED GRID project. She is also an Associate Professor with the Electrical Engineering Department at JSS Academy of Technical Education, India. She is a Senior Member of IEEE, Life Member of ISTE, IETE, IEI, and IAENG. She has published over 70 research/review papers in IEEE Journals and has registered seven patents.

R. Sharmila is an Assistant Professor in the Department of Computer Science Engineering at Krishna Engineering College, India. Her research is highly interdisciplinary, spanning a wide range of topics, including Wireless Sensor Networks, Digital image Processing, Cryptography and Information Security, and Blockchain.

Contributors

Sharmistha Adhikari
Department of Computer Science
and Engineering
National Institute of Technology
Sikkim, India

Pastor Arguelles
College of Computer Studies
University of Perpetual Help
System DALTA
Calabarzon, Philippines

Navneet Arora
Department of Computer
Science
University of Liverpool
Liverpool, England

Alaknanda Ashok
Department of Electrical
Engineering
GBPUAT
Uttarakhand, India

Bharat Bhardwaj
Department of IT
AKTU
Lucknow, India

Piyush Dhuliya
Department of Electronics and
Communication Engineering
Tula's Institute
Dehradun, India

Priyanka Dhuliya
Department of Electronics
Engineering
Graphic Era Hill University
Dehradun, India

Asmita Dixit
Department of IT
ABES Engineering College
Ghaziabad, India

Hoor Fatima
Department of Computer Science
and Engineering
SET, Sharda University
Greater Noida, India

Ruchi Goel
Department of Computer Science
and Engineering
AKTU
Lucknow, India

Kanika Gupta
Department of IT
ABES Engineering College
Ghaziabad, India

Shruti Gupta
Department of CSE
National Institute of Technology
Delhi, India

Aatif Jamshed
Department of IT
ABES Engineering College
Ghaziabad, India

Jasmeet Kalra
Department of Mechanical
  Engineering
Graphic Era Hill University
Dehradun, India

Umang Kant
Department of Computer Science
  and Engineering
Delhi Technological University
Delhi, India
and
Department of Computer Science
  and Engineering
Krishna Engineering College
Ghaziabad, India

Ashish Kumar
Department of CSE
Krishna Engineering College
Ghaziabad, India

Dhananjay Kumar
Department of Computer Science
  and Engineering
National Institute of Technology
Sikkim, India

Dinesh Kumar
OSI Soft Australia Pvt. Ltd.
Perth, WA, Australia

Manish Kumar
Department of CSE
Krishna Engineering College
Ghaziabad, India

Pramod Kumar
Department of CSE
Krishna Engineering College
Ghaziabad, India

Raja Kumar
School of Computer Science &
  Engineering
Head of Research - Faculty of
  Innovation and Technology
Taylor's University
Malaysia

Vinod Kumar
Department of Computer Science
  and Engineering
Delhi Technological University
Delhi, India

Abirami Manoharan
Department of Electrical and
 Electronics Engineering
Government College of
 Engineering
Srirangam, India

Hariprasath Manoharan
Department of Electronics and
 Communication Engineering
Panimalar Institute of Technology,
 Chennai, India

Prashant Naresh
Department of Computer Science
 and Engineering
AKTU
Lucknow, India

Pankaj Negi
Department of Mechanical
 Engineering
Graphic Era Hill University
Dehradun, India

Diwaker Pant
Department of Electronics and
 Communication Engineering
Tula's Institute
Dehradun, India

Rajesh Pant
Department of Mechanical
 Engineering
Graphic Era Hill University
Dehradun, India

Nitin Rakesh
Department of Computer Science
 and Engineering
SET, Sharda University,
Greater Noida, India

Sangram Ray
Department of Computer Science
 and Engineering
National Institute of Technology
Sikkim, India

Pavi Saraswat
Department of Computer Science
 and Engineering
AKTU
Lucknow, India

Sunil Semwal
Department of Electronics and
 Communication Engineering
Tula's Institute
Dehradun, India

Shankar T.
Department of Electronics and
 Communication Engineering
Government College of
 Engineering
Srirangam, India

Yogesh Kumar Sharma
Department of Computer Science
 and Engineering
AKTU
Lucknow, India

R. Sharmila
Department of Computer Science
and Engineering
Krishna Engineering College
Ghaziabad, India

Meenu Shukla
Department of Computer Science
and Engineering
Krishna Engineering College
Ghaziabad, India

Amit Sinha
Department of IT
ABES Engineering College
Ghaziabad, India

Sandeep Tiwari
Department of Mechanical
Engineering
Krishna Engineering College
Ghaziabad, India

Anuradha Tomar
Department of Instrumentation &
Control Engineering
Netaji Subhas University of
Technology, Govt. of NCT of
Delhi
Delhi, India

Neha Tyagi
Department of Computer Science
and Engineering
SET, Sharda University
Greater Noida, India

Yuvaraja T.
MOBI-Mobility, Logistics and
Automotive Technology
Research Centre
Vrije Universiteit Brussel
Ixelles, Brussels, Belgium

Jawwad Zaidi
Dr. Akhilesh Das Gupta Institute
of Technology & Management
Delhi, India

Fatima Ziya
Department of Computer Science
and Engineering
Krishna Engineering College
Ghaziabad, India

CHAPTER 1

Introduction to Emerging Technologies in Computer Science and Its Applications

Umang Kant and Vinod Kumar

Delhi Technological University

CONTENTS

DOI: 10.1201/9781003121466-1

1.1 INTRODUCTION

The extensive and exhaustive research carried out in the field of Artificial Intelligence (AI) is a confirmation that it finds its applications in every field of life now-a-days. Researchers and scientists are making every possible effort to help the world by using AI, in turn making machines which think and maybe act like humans. We are aware that AI is like an umbrella that shelters numerous technologies under it and hence it is perceived as an interdisciplinary field with several approaches. AI is an eclectic branch of Computer Science that aims to respond to Turing's question in assenting and is responsible for developing smart machines capable of executing tasks that require human intelligence and is responsible for a visible paradigm shift in every sector of the technological world and thereby giving birth to new concepts and technologies on the way. Machine Learning is an application of AI, which aims at offering the machines or systems the capability to learn on its own and improve its experiences at every turn without human intervention. In order to make the machines learn, we need to provide them with ample amount of data so that machines can analyze some pattern in the data (if any) and make better decisions based on observing the data, working on the patterns of the data and then training the algorithms using that data [1,2]. Learning process is initiated by observing the data as mentioned above, and learning techniques can be as follows: supervised, unsupervised, semi-supervised, or reinforced based on the data to be trained and the application to be addressed. Hence, the main aim of Machine Learning is to allow the machines to learn on their own in the absence of human assistance and adjust their output or actions accordingly [3]. AI has given birth to many new technologies and Machine Learning is one of the ways to achieve AI. We will now be discussing some recent technologies which have been researched the most these days and are finding their way in every aspect of business, education, health, commercial, and other fields of life.

1.1.1 Computer Vision

Computer vision is such a type of AI which we all have naturally experienced in our lives in multiple ways without even realizing it. Such is the power of human brain and senses. Computer vision aims to replicate this power of human brain and senses using machines. Humans can (i) describe the content of the image or video they have seen, (ii) summarize an image or video they have seen, and (iii) recognize a face or an object that they have seen [4]. Hence, a machine can take advantage of human's capa bility of remembering things and people they have seen and make their machines learn the same capability using algorithms dedicated for this process. We all are aware that taking and uploading a picture or video on internet has become extremely easy; all we need is a smartphone and some social media platform. According to recent articles, around hundreds of videos are uploaded per minute on YouTube social platform; and the same is the case with other social platforms such as Facebook, Instagram, and Twitter. Around 3 billion images alone are shared online each day, maybe more. These images and videos can be easily recognized and summarized by humans but to train the machine for the same capability, we first need the machine to be able to index the content and then be able to search the content in that video or photograph. The Machine Learning algorithms will need to (i) find what the image or video contains and (ii) utilize the metadata descriptions provided by the person who has uploaded that image or video.

In simple terms, computer vision can be defined as a field of study focused on the problem of helping computers to see [5]. The goal of com puter vision is to use the observed image data to infer something about the world [6]. Computer vision is an interdisciplinary technological field which deals with replicating and observing the human vision and brain processing system and facilitating the machines to identify and process items in images and videos in a similar manner as humans are capable of. Due to the advancements in AI, Neural Networks (NNs), and Deep Learning, computer vision has taken great leaps in recent years and is still a hot field among researchers [7]. Computer vision is also clearly a sub field of AI, Machine Learning, and Deep Learning, as it deals with overly complex data identification and interpretation. Due to recent advance ments, computer vision has been able to successfully outdo humans in tasks of identifying, indexing, and labeling objects in the images or vid eos. This must have been experienced by many users while tagging people in images using social media platforms such as Facebook. The algorithms are trained in such an extensive manner that now they can perform bet ter than humans in identifying and tagging items or people [8]. Another factor which is responsible for the better working of machines to achieve computer vision is that over the past few years, ample amount of data is being generated these days. Large amount of data generated is being used for training of Machine Learning algorithms which, in turn, leads to better results. The concept and aim of computer vision can be understood by referring to Figure 1.1.

Research and experiments on computer vision can be traced back to 1950s. But it was used commercially in 1970s to distinguish between handwritten text and typed text. Back then, there was not as much abundance of data as is the case now. Today, the applications have also increased as there is a huge amount of data, better trained algorithms and hardware. The data is so huge that these days data scientists are being hired to just analyze and work upon the data. It is the job of the data scientists to filter the data and select the data sets to be used for training of the algorithms. The computer vision market is expected to reach $48.6 billion by early 2022 [5].

Another topic of discussion is how does computer vision work? This question is not simple to answer as computer vision is inspired by the human brain. Since it is still not clear how human brain and eyes work to interpret the objects, it is difficult to approximate the same concept in algorithms. All we can say is that computer vision mimics the way human brain works. However, it is difficult to comment on how well the developed algorithms will be able to copy the human brain and implement it

FIGURE 1.1 Object detection and classification. (From Jarvis, R.A., *IEEE Transactions on Pattern Analysis and Machine Intelligence*, 122–139, 1983. With permission.) [5].

on machines. Current algorithms are heavily using Pattern Recognition approach to achieve computer vision. Hence, to make the computer understand and to train the algorithm, the approach used is to feed the system with a lot of data (i.e. images of the objects to be identified). Larger the data set, better labeling would be achieved. Then, these labeled images or objects would be subjected to various algorithms or software to identify the patterns with the objects and thereby in classification (Figure 1.1). When a large data set of images is fed to the machine, the algorithms analyze the angles, shapes, colors, and borders of the objects and the distance between various objects. After this analysis, the machine would match these unlabeled images with the labeled images and put them in the respective image set of objects (single object image or multiple object image). The basic working of computer vision algorithm is divided into (i) classification, (ii) localization, (iii) object detection, and (iv) instance segmentation. Algorithms train better when the input image data set is big, even though large image data set means larger memory requirements, but in the end, we get a better trained algorithm.

In recent years, Machine Learning and Deep Learning technology produce better results as compared to previous efforts where the working of machines was extremely limited. Earlier, a set manual approach was used where first a database was created in which all captured images were stored that needed analysis. Second, all images had to be annotated, i.e. many key points were added to the images for labeling them, and manually rules were coded for each object in the image. Third, new images were captured, and the entire process had to be repeated. When Machine Learning came into the picture, the manual rule coding was eliminated and feature extraction using various Machine Learning algorithms such as Support Vector Machine (SVM) and linear regression were used to identify patterns and classify further. However, in recent years, Deep Learning has given a new approach to apply Machine Learning to the data set by using NNs. The detailed working of Deep Learning for computer vision is out of the scope of this chapter.

Computer vision finds many applications today and users are making use of computer vision knowingly or unknowingly. The applications of computer vision include (i) computer vision in facial recognition, (ii) computer vision in self-driving cars, (iii) computer vision in healthcare, and (iv) computer vision in augmented reality (AR).

1.1.2 Deep Learning

Deep Learning is a subset of Machine Learning that aims to further automate the functions of a human being.

It is a branch based on the building algorithms that learn and re-learn by mimicking the functions of a human brain.

Just like the NN helps humans learn from their experiences, artificial neural networks (ANNs) help an algorithm learn and execute the task.

ANNs, also generally called NNs, are computing systems vaguely inspired by the biological NNs that constitute animal brains.

An ANN comprises multiple artificial neurons (or nodes) arranged in a network of multiple layers, which loosely models the NN of the biological brain (Figure 1.2).

Each connection, like the synapses in a biological brain, can transmit a signal to other neurons.

The "signal" comprises input data (in real numbers) and then processes it before sending it further down the chain. Every neuron processes the data before transmitting it further.

Different layers of the neurons perform different transformations on their inputs. Signals travel from the first layer (the input layer) to the last layer (the output layer), possibly after traversing the layers multiple times.

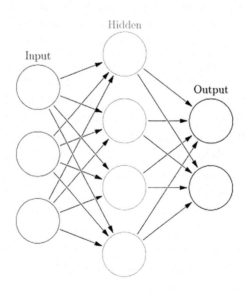

FIGURE 1.2   Artificial neural network layers.

There are a few different types of ANNs designed to execute different tasks. The major ones are listed below.

- Multilayer Perceptron Networks.

- Convolutional NNs.

- Long Short-Term Memory or Recurrent NNs.

Once the data is prepared for processing, it is fed to the NN in the first layer, also known as the input layer.

Now the neurons process this data and send it further down to the next layer. Each layer is designed to perform a specific task. The middle layers are also called the hidden layers.

Once the data is sequentially processed by each hidden layer, the model transmits the processed data to the output layer (the last layer).

In the output layer, the data is further processed finally as designed and gives out the final model. This one pass of the data through the network is called an epoch.

This output is then tested for accuracy; most often than not, the output of the first epoch is far from being correct. Therefore, this information is passed back to the network in reverse order so that the network can learn from its mistakes. This is called back-propagation.

The network then tweaks its parameters further and processes the data in a similar fashion. This process is continued for several epochs until the model starts producing acceptably accurate results.

The theory, model, and the data existed earlier as well, though it is only the advancements in the technology that have empowered this vision into reality.

Today we have access to sophisticated data management architectures and the computational power to process this massive data. This has made the access to these technologies fairly simple.

The most prominent technologies on this front are TensorFlow, Keras, and Pytorch that have enabled everyone to access the state-of-art technology of Deep Learning.

There are not many differences between Machine Learning and Deep Learning. Here are a few.

While Machine Learning is based on pre-defined models or algorithms, Deep Learning is built on NN architecture. It further removes the need for human intervention for feature selection in the data.

Since Deep Learning is state-of-the-art, it requires high computational power to be processed, which is now possible with the advanced GPUs.

Deep Learning finds its applications in several industries. Some of the major applications of Deep Learning are as follows:

- Self-driving cars

- News aggregation and fraud news detection

- Natural language processing

- Virtual assistants

- Recommender systems

- Visual recognition

- Credit fraud detection

- Healthcare and diagnostics

There are many more. The possibilities are endless (Figure 1.3).

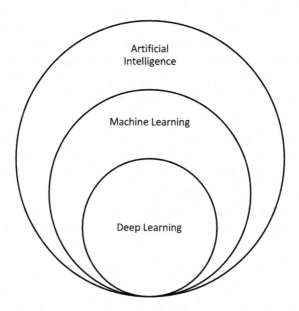

FIGURE 1.3 Relationship between artificial intelligence, machine learning, and deep learning.

1.1.3 Internet of Things (IoT)

Researchers describe Internet of Things (IoT) as a wide network of physical devices or objects connected with each other over the internet for the purpose of exchanging data and communication with each other, as depicted in Figure 1.4. These devices can also be called as smart devices as they require minimum or no human intervention for their processing. These physical devices are embedded with sensors and technologies for establishing connection with other devices. Today such devices range from basic household devices to much sophisticated engineering devices. As predicted by experts, the number of connected IoT devices is expected to increase from current 8 billion to 45 billion by 2025 or maybe more [9]. Examples of such devices range from a smart coffee maker, smart lightbulb, smart toy, baby monitor, and driverless vehicle to a smart city. A device being controlled by a smartphone or a controlled environment fall under the category of IoT or smart devices. IoT has become one of the most sought after technologies of recent times to provide effortless and continuous communication between people and things (devices) [10]. This has made every possible device connect to other devices over the internet via embedded devices. These devices cooperate by making the physical world meet the digital world [11].

FIGURE 1.4   Device connectivity over the internet.

The connection of a large number of embedded devices over the internet has been possible due to the availability of mobile technologies, low-power sensor technologies, low-cost computing capabilities, cloud connectivity, Machine Learning and big data analytics, and AI. These devices communicate, share, and collect data with other embedded devices with minimum human intervention; hence, the demand of IoT devices is increasing gradually. IoT has large business scope due to its benefits in marketplace. Benefits include: (i) new IoT-based business models, (ii) increased productivity due to smart connectivity, (iii) efficient business operations, (iv) better revenue modes, and (v) smooth connection between physical business world and digital world [12]. Due to these benefits, IoT has become widespread in the industrial world and there is a new classification of IoT: Industrial Internet of Things (IIoT). IIoT refers to the merging of the IoT technology in industrial environments. To integrate IoT in industrial settings, industries have used cloud, analytics, Machine Learning and machine-to-machine (M2M) communication for wireless automation and control among connected devices.

This integration of IoT in industries has led to a large number of application areas such as smart homes, smart cities, smart logistics, smart manufacturing, smart power grids, smart digital supply chains, and predictive logistics and maintenance, among others.

Devices have been able to connect digitally over the internet low-power chips and cloud technologies, and the device connectivity will keep on increasing as discussed above. However, this device connectivity is highly vulnerable to security threats as these devices collect and communicate almost all of the user's sensitive data over the internet. Until now, security has been the biggest area of concern and the most researched area of IoT. Researchers are continuously working on to make the devices self-capable in order to find the patches of the detected faults in security. These software faults are regularly discovered and hence their patches are of utmost requirement, but many devices lack the capability to remove the risk factors of these faults, thereby keeping the devices and users at permanent risk, ultimately making the user vulnerable to hackers. Hackers can hack the sensors of the device and control the entire environment by manipulating the sensors without the user's knowledge, and this threat can be minor or catastrophic. One example of such security vulnerability in users wearing a smartwatch is hackers being able to eavesdrop the user's conversations even when the microphone is switched off or track the user's location all the time even when internet is disabled. And such security threats put all users at risks in using small household devices and business or industrial devices. As discussed above, IoT connects physical devices with digital devices; hence, the real world between the two is always at risk. Hence, the current area of research in IoT these days revolves mostly around security concerns.

As security is a major concern in IoT, so is privacy. By always being connected over the internet using embedded devices, privacy takes a back seat in the entire process and hence the security lapse. IoT models have user's data which can be manipulated to achieve some cause or worse it can be sold to companies or can be made available over the dark web. Hence, it becomes equally important for the users to be aware about the bargain they make while using these smart devices. Apart from security and privacy, other IoT concerns can be cost, connectivity, user's acceptance, and device standards.

IoT has given rise to big data analytics as IoT generates vast amount of data on a daily basis and hence has given companies vast data sets to analyze and work upon. This data can be in many different forms such as images, videos, audios, pressure or temperature readings or heartbeat, or other sensor readings. This vast amount of data has given rise to metadata, which contains data about data. This huge data cannot be stored in the company's data warehouse or other resources, but on the cloud. Hence, IoT, in turn, has given rise to the need for cloud services with every company aiming to achieve the IoT business model. These days, from small organizations and institutes to large multinational companies, all are making use of cloud services to better manage the data. With better connectivity (3G, 4G, 5G, CDMA, GSM, and LTE networks) and new technologies, the IoT market will continue to evolve even with security and privacy issues.

1.1.4 Quantum Computing

Current generation computers are based on classical physics and therefore on classical computing. In theory, i.e. Turing machines and in practice, i.e. PCs, laptops, smartphones, tablets all are current generation computers and work on the principle of classical computing and are called classical computers. These classical computers can only be in one state at a particular time and their operations can only have local effects, i.e. they are limited by locality [13]. As we are aware, the fundamental principle behind the computer systems is the ability to store, fetch, and manipulate information. This information is stored in the form of bits, i.e. binary 0 and 1 states, and all the manipulation is carried out by using these binary 0 and 1 states only [14]. These two bits are used in all classical computers. However, we have shifted from classical physics to quantum physics, as the real world behaves quite randomly and this behavior is not fixed; and to capture this random behavior of world using the machines, we cannot always use classical computers, or computers using classical computing as they remain in a single state at a particular time as discussed above, hence the shift from classical computers to quantum computers. A quantum computer can be in different states at the same time and the different states can be superimposed upon each other, and hence interference among different states can be achieved during processing. This superimposition of states can be achieved by using quantum bits, also called as qubits [14]. Hence, quantum computers manipulate the information stored by applying quantum mechanical principle using qubits instead of bits. A cluster of these quantum computers [15] can be either locally placed or spatially distributed but, in any case, they can achieve the non-local effects due to state superimposition.

As discussed above, quantum computing uses the principles of quantum mechanics and examines the processing power and related properties of computers. The main aim of quantum computing is to develop quantum algorithms to be used in quantum computers which would be faster and better than classical algorithms being used in classical computers [15]. Quantum computing is aimed to speed up the computation and to solve computationally hard problems which are not only difficult but next to impossible to solve using classical computers.

The field of quantum computing started with the use of analog quantum computers in the early 1980s by Yuri Manin [16], Richard Feynman [17,18], and Paul Benioff [19]. Although the work in this field has not been progressing at an extremely fast pace, the first noted algorithms in this field were developed later by Deutsch [20] and Simon [21]. The work on quantum complexity theory was initiated by Bernstein and Vazirani [14]. The work pace and interest in the field caught up fast in 1994 when Peter Shor discovered the efficient quantum algorithms for integer factorization and discrete logarithms problems [20].

Quantum mechanical phenomena use three basic quantum properties: (i) superposition: it is a condition where two or more independent states are combined to yield a new state. An example of such superposition would be the combination of two or more music notes, and the final music that we hear would be the super positioned note; (ii) Entanglement: it is a counter-intuitive quantum behavior which is not visible in the classical environment. Here, the object particles are entangled with each other and form a new model or environment. This new model behaves in entirely new ways which is not possible in the classical world and also cannot be explained using classical computing logic; and (iii) Interference: interference of quantum states occurs due to the logic of phase. Due to the phenomenon of phase, the states undergo interference. This logic of state interference is similar to the logic of wave interference. In wave interference, the wave amplitudes add when they are in phase, else their amplitudes cancel each other. To develop a fault-tolerant quantum system and to enhance the computational capabilities of a quantum computer, researchers are working toward improving the (i) qubit count: the aim to create more qubit states; more the qubit states, more the options of manipulation and processing of states; and (ii) low error rates: the aim is to eliminate the possible noise and errors encountered while working on multiple qubit states. Low error rates are required to manage qubit states in an efficient manner and to perform all sequential or parallel operations. Volume is considered to be a useful metric for analyzing quantum computer capability [14]. Volume measures the correlation between the quality and number of qubits, the error rates of qubit processing, and the quantum computer circuit connectivity. Hence, the aim is to develop quantum computers with large quantum volume for solving computational hard problems [22,23].

The basic motivation to research in the field of quantum computing is that the classical computers themselves have become so powerful and cheap due to miniaturization that they have almost already reached micro levels where quantum processing and effects are said to occur. The chip makers have led to such a level of miniaturization that instead of suppressing the quantum effects in classical computers, researchers might try to work with them, leading to further miniaturization and hence more quantum effects. It is too soon to comment on the pressing question "to what extent will quantum computers be built?" The first 2-qubit quantum computer was developed in 1997, and then a 5-qubit quantum computer was built in 2001 to find the factor of number 15 [21]. The largest quantum computer up until now has only few dozen qubits. Hence, the work on quantum computers has been rather slow but at a steady pace.

1.1.5 Edge Computing

With the rise in IoT-connected devices, edge computing has also come into the picture and is transmuting the manner in which data is being processed, managed, stored, and distributed to the users by millions and millions of connected devices around the globe. As discussed in Section 1.1.3, in IoT, the connected devices generate tremendous amount of data which is stored and retrieved from either a centralized storage location or cloud-based storage location. And this data is expected to continue to grow at an unprecedented growth. Hence, more time is being spent in storing and retrieving the generated data. IoT, real-time computing power, and faster networking medium and technologies like 5G (wireless) have aided edge computing with a large number of opportunities in business industries.

As per the definition given by celebrated researcher Gartner, edge computing is a part of a distributed computing topology in which information processing is located closer to the edge, also called as edge nodes (Figure 1.5), where things and people produce or consume that information [24]. It is understood that edge computing gets the data storage and its processing closer to where the data is being generated. Hence, edge computing is a decentralized distributed computing framework bringing the enterprise data closer to data sources [25].

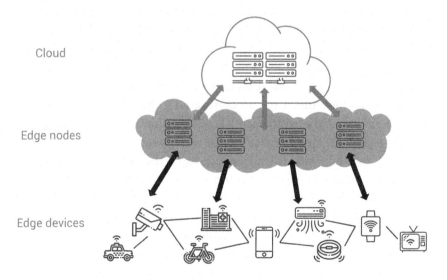

FIGURE 1.5   Edge nodes and edge devices.

This, in turn, helps improve bandwidth availability, data control, better insights, swift and appropriate actions, reduce costs, comprehensive data analysis, and continuous processing. Another advantage of edge computing has come in the form that the maximum amount of data generated can be optimized for processing and manipulation, whereas currently even though a large amount of data is being generated and stored, majority of it is not being used for processing due to latency and network issues. Due to edge computing, since the maximum data is to be utilized for processing, i.e. maximum data potential is extracted, better insights and better analysis (predictive) are achieved which, in turn, improves the overall quality [25]. This contrasts with the current computing where the data must be stored at a central location which can be far away from where the data is being generated. Hence, here the need for centralized storage location or cloud-based storage location is eliminated. This concept of storing and processing data closer to the source is introduced to reduce the problem of (i) bandwidth issues, (ii) real-time data latency which affects the overall performance of the software or application, as in edge computing, the data does not travel over a network to a central storage location or cloud for further processing, and (iii) this processing environment helps organizations save overall cost. As per the current studies, almost 75% industry data will be processed at the edge by 2025 [26]. Currently, only around 8%–10% data is being processed at the edge. At the moment and also in the past, extraordinary volumes of data generated by IoT-connected devices have led to the outperformance of the current infrastructure and network capabilities. Hence, with the promise of upcoming, faster networking technologies such as 5G, edge computing gives a solution. The advent of edge computing has facilitated the work of real-time applications such as smart homes, self-driving cars, robotics, animation, and video processing, among many others.

The technology of edge computing also brings a few challenges along with its advantages. Issues which must be addressed are (i) bandwidth limitations, (ii) latency issues, (iii) data complexity management, and (iv) network security and risks, among others. An effective edge computing model must also be able to (i) seamlessly deploy applications to edge locations, (ii) manage data workloads across centralized storage and cloud locations, (iii) address all security concerns, and (iv) sustain flexibility to evolve according to the environment.

Edge computing though different from fog computing finds many similarities with it. This concept is discussed in the following section.

1.1.6 Fog Computing

Fog computing as defined by researchers is a decentralized, distributed computing technology where the data, storage resources, computation technology, and respective applications are located at someplace between the data source (IoT-connected devices) and the cloud [27]. Fog computing finds many similarities with edge computing as both bring the data source and computing near to the location where the data is being created (see Section 1.1.5). Fog computing also finds similar advantages to edge computing such as it also reduces data latency and provides better efficiency among other advantages. Hence, many scholars interchangeably use both terms, as the basic aim of both technologies is same. Although the main motive and working are similar in both, there still exists minor differences between the two. While edge computing implies to getting the data storage and processing closer to data sources, i.e. edge nodes, fog computing refers to getting the data storage and processing in-between the data source and the cloud, i.e. fog nodes [28]. As can be understood by the name, in nature, fog concentrates between the ground and the cloud, to be precise it stays in-between but still closer to the cloud. Hence, the term fog computing has been coined by Cisco in 2015 by the company's product line manager, Ginny Nichols [29]. Fog and edge computing can be viewed as two sides of a coin, as both complement each other and both have similar advantages and limitations. The same can be understood by referring to Figure 1.6.

Fog computing along with edge computing is viewed as an alternative to cloud computing. Both retain few properties of cloud computing, at the same time maintaining few distinctions [30]. Smart electrical grid, smart transportation networks, and smart cities are all applications of fog computing.

The implementation of fog computing requires bringing IoT applications at the fog nodes at network edge using fog computing models and tools. Here, the fog nodes, which are closest to the edge nodes, receive the data from other edge devices (modems, routers, sensors etc.). After receiving the data, it is then forwarded to the optimum location for further analysis. In fog computing, the data analysis is based upon the type of data. The data can be divided into different categories based on the

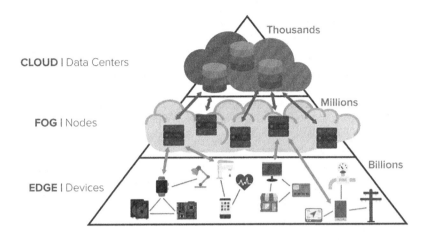

FIGURE 1.6 Representation of cloud, fog, and edge computing. (From Gupta, B.B., & Agrawal, D.P. (Eds.), *Handbook of Research on Cloud Computing and Big Data Applications in IoT*. IGI Global, 2019. With permission.) [31].

time sensitivity. The data which needs immediate analysis and processing can be put into most time-sensitive category, i.e. the user needs this type of data immediately, and the categories can hence vary from most time-sensitive to medium time-sensitive to least time-sensitive data. Among the collected data set, the most time-sensitive data is selected and is analyzed as near to its source as possible to avoid latency issues. The data which is not time sensitive is sent to aggregation nodes for analysis, which can be carried out at the approximate time as the user does not require it immediately. The benefits of fog computing, which are similar to edge computing, are (i) reduced latency, (ii) better network bandwidth, (iii) improved reliability, (iv) reduced costs, and (v) better insights, among others. The differences are listed in Table 1.1.

1.1.7 Serverless Computing

Serverless computing is an auto-scaling computing because here the company or the organization gets backend services from a serverless vendor [32]. In serverless computing, the backend services are provided based on user's requirements (event-driven), and hence the users do not need to worry about the underlying architecture and infrastructure and just need to implement the codes and further processing without any conundrum. An organization taking services from a serverless retailer is only charged for the extent of the processing carried out by them and they need not pay

TABLE 1.1   Difference between Fog Computing and Edge Computing

S. No.	Fog Computing	Edge Computing
1.	In fog computing, the fog nodes are responsible for deciding whether the data generated from various smart IoT devices is to be processed using its own resources or has to be sent to the cloud for storing and computing.	In edge computing, each edge node manages, stores, and processes the data right at the edge of the source, i.e. locally instead of transferring it to the cloud.
2.	Fog computing distributes the storage, communication, and further processing of data close to the cloud keeping it in control of end user, rather than exactly close to the source of data.	Edge computing locates the storage and other processing close to the source of data. Here also, the control is in the hands of end users.
3.	Fog computing manages the intelligence down to the level of local area network of the network architecture.	Edge computing manages the intelligence of the edge gateways into the devices generating the data, thereby keeping the entire computing close to the devices.
4.	In fog computing, fog nodes work with the cloud.	In edge computing, the cloud involvement is eliminated.
5.	Fog computing forms a hierarchical-layered network.	Edge computing is limited to individual edge nodes which do not form a network.

a fixed fee for the server architecture, number of servers, and the bandwidth [33]. The price to be paid scales based on the services provided by the serverless retailer to the organization; hence, serverless computing is also called as auto-scaling computing. Serverless retailers provide storage and database services to the users [34].

As discussed, the users are not aware about the server architecture and the number of servers used. This does not mean that the environment is free from servers; physical servers are present and are used but the users are not aware about their existence. Since the underlying architecture is not considered by the user, this serverless environment becomes cost-effective and more user-friendly. The cost-effectiveness of this computing can be referred to in Figure 1.7. Apart from being cost-effective, serverless computing has other benefits as well: (i) fully managed service, (ii) dynamic scalability and flexibility, (iii) only pay for the resources or services used (as discussed above), (iv) enhanced productivity, (v) flawless connections, (vi) better turnaround time, (vii) no infrastructure management, and (viii) efficient use of resources, and others.

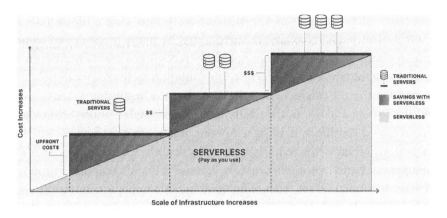

FIGURE 1.7 Cost-effectiveness of serverless computing.

Although having many advantages, serverless suffers from few disadvantages too. One of the major limitations of serverless is cold start. Cold start is a condition when the latency increases significantly. When a certain serverless function has not been used or called by any user for some time, the serverless provider closes down that function to save space and energy, making that function go dormant. Whenever any user calls for that closed-down function, the serverless vendor will have to again start and host the function from the scratch, and this entire process takes significant amount of time, which adds to the latency. This condition is called as cold start. As opposed to cold starts, the functions which are being frequently used by the users correspond to warm starts.

1.1.8 Implanted Technology

The implanted technology is a type of AI through which we can enhance the capability of the human body organs with the help of computer vision [35]. As discussed in Section 1.1.1, computer vision is a technology that will replicate the power of human brain and senses in a machine, and with the help of that machine we can enhance the power of our own senses. The main question arises why the need of implantation technology. In order to find the answer to this question, we need to look at the uses of implantation technology first. The need of implantation technology can be better analyzed by understanding a few examples where it can be used: (i) humans do not have a memory that will store that data forever, similar to what a computer hard disk does, so here we can use brain implantation technology that will make our memory permanent and we will remember each and every moment and experiences of our life from birth until death, (ii) in today's environment we see that we are surrounded by many diseases and some diseases such as cancer that are not easily recognized and hence become fatal if not recognized earlier. Here, implantation technology can help us in this way that we can recognize the disease at the beginning so that it can be cured soon; and (iii) in the future, we can use implantation technology in our day-to-day life by making use of our organs as smart devices, which, in turn, will remove the requirement of physical electronic devices such as smartphones and laptops. because we can use AR and computer vision to achieve it. Implantation technology uses brain computer interface (BCI) technology which, in turn, is also a heavily researched interdisciplinary area. The layout of the BCI is shown in Figure 1.8.

BCI is a technology with which we can provide instructions to the machine through the input signals of the brain. We all know that our brain is made up of neurons, as shown in Figure 1.9, and these neurons have electrical signals (impulses) in them which are generated when a person thinks or feels anything. The BCI technology will sense those signals and convert them to computer readable format (binary format). There is a subset of AI known as Machine Learning that focuses on training the machine by feeding the data to produce some useful productions and there is a subset of Machine Learning known as Deep Learning that mainly focuses on replication of neurons to machine through a NN, as shown in Figure 1.10 (both Machine Learning and Deep Learning are discussed in Sections 1.1 and 1.1.2 of this chapter). A NN is a mathematical modeling of a neuron of our brain and with the help of this network, we try to replicate the functionality of the brain into our machine.

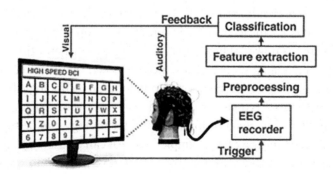

FIGURE 1.8   Basic BCI diagram.

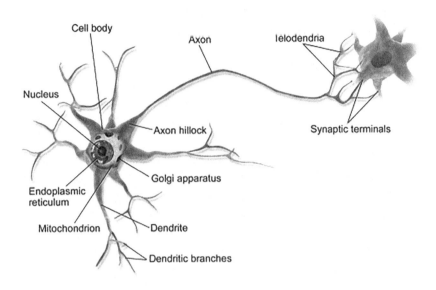

FIGURE 1.9   A neuron.

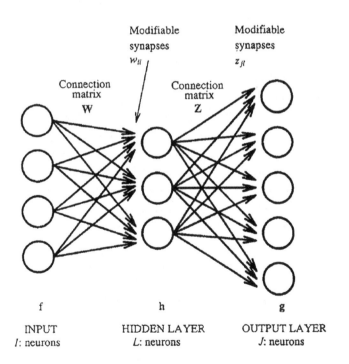

FIGURE 1.10   A neural network. (From Abdi, H., *Journal of Biological Systems*, 2(03), 247–281, 1994. With permission.) [36].

Our neurons work on action potential, and the propagation (Figure 1.11) of this action potential makes us realize that with implantation technology, we try to sense this action potential and generate a machine level code from it through which the machine will be able to understand the action that is needed to be performed. Notwithstanding the shortcomings and hurdles, with the activity potential component effectively surely known, further advances in microelectronic advances empowered improvement of the neural probe, as shown in Figure 1.12. Recent studies have revealed improvements and progressed the possibility of a neural test by presenting the idea of "neural dust," an enormous number of remote cathodes that can be appended straightforwardly to various nerves consequently making numerous remote detecting hubs inside the body.

Anticipating what technology is to come in future is an assignment that engineers generally leave to futurists and sci-fi authors and movie makers. Nevertheless, as scientists and researchers, the whole science community takes motivation from such futurists and work towards a common goal by trying to benefit from these ideas, which invigorates scholarly discussions and research so that all potential situations may unfurl later on. At the

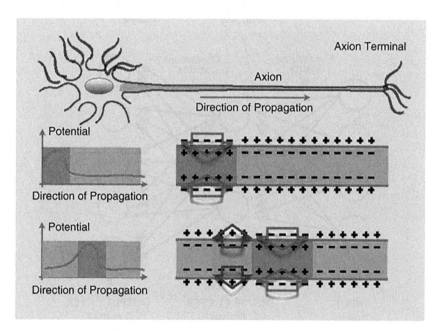

FIGURE 1.11　Action potential and its flow. (From Sobot, R., *IEEE Technology and Society Magazine*, 37, 35–45, 2018. With permission.) [37].

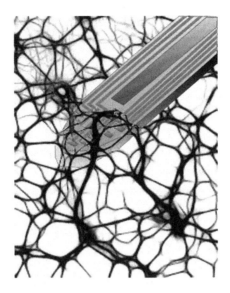

FIGURE 1.12 A neural probe. (From Sobot, R., *IEEE Technology and Society Magazine*, 37, 35–45, 2018. With permission.) [37].

moment, we observe this technology mostly in theory either in readings or watching these possibilities in science fiction works. However, as time progresses, advances in Machine Learning that mimic living processes, coupled with the developments in self-healing materials and development of bionic skin, further deepen the prospect of arriving at a fully bionic post-human.

1.1.9 Virtual, Augmented, and Mixed Reality

The term virtual means proximity and the term reality refers to experiences as a human being. If we put these two terms together, it becomes virtual reality (VR) and the meaning becomes an experience which is close or near to reality. VR is simulation of the real-world phenomenon [38]. This VR is presented to human beings by bringing their sensory organs in the presence of such a version of reality which does not exist. But our sensory mechanism and brain perceive it as something real. Our sensory mechanism and brain are projected to such amusing collections of made up information that our brain perceives it as real, even though the perception is not actually present. In a more technical form, VR is a representation of a 3D system-created environment which interacts with a user in such a way that the user becomes a part of it and is completely immersed in it. It becomes difficult for the user to distinguish the virtual world from the real world and hence take decisions or actions based on the virtual information being presented. The VR environment can manipulate the user with the virtual information. Although it sounds simple, our brain is so complex and evolved that even though confused for some time can ultimately tell the difference between the virtual world and the real world. However, during that moment of illusion, the user can be manipulated by the system. To achieve this VR, the environment has a set of systems and devices which are used by the users to experience the virtual world. These devices can be gloves, headsets, glasses, etc. These devices emulate our sensory organs, i.e. our senses and generate an impression of reality [39]. The need of VR is simply to engage and attract users and customers to a particular field. Also, VR is used whenever there is a scope of large expenses, danger, or impracticality. VR changes the perception of the users just by presenting a particular object in the 3D format. VR is currently being used in various applications, including (i) entertainment, (ii) sports, (iii) medicine, (iv) architecture, and (v) the arts [40]. Figure 1.13 depicts few examples of VR environments.

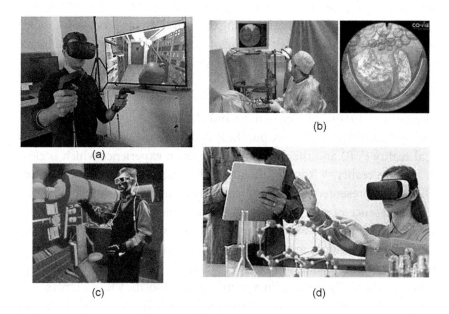

FIGURE 1.13 (a) VR in gaming (from European Space Agency), (b) VR in medicine [41], (c) VR in automation [42], and (d) VR in education [43].

One of the latest and biggest technology trends is AR, which is a variation of virtual environments (VE) or VR [44]. Although the concept is old, the usage has been quite recent. As discussed above, the VE technologies immerse the users completely inside this virtual environment, where the user becomes so engrossed that he/she becomes a part of that virtual environment and starts interpreting this virtual world as the real world. The vehicle testing simulation environments can be classic examples of VR or VE technologies. The variation comes in the form of AR [45,46], where the user can make the difference between the virtual world and the real world, i.e. the virtual objects are superimposed on the real-world objects. Hence, AR complements reality, rather than entirely replacing it. An example of AR can be a game named *Pokémon Go*, where the players can locate, capture, and play with *Pokémon* characters that turn up in the real world and real objects, such as parks, subways, bathrooms, lawns, and roofs. This game had become a rage and had to be controlled and banned in many parts of the world. Apart from games, AR is being used in many other fields as well: (i) news broadcasting, where the news presenters or anchors can draw lines and other shapes on the screen; (ii) navigation systems, where the routes are superimposed on the actual roads; (iii) defense, where the military personnel can view their status and positionings on their helmets; (iv) medicine, sometimes neurosurgeons use AR projections of a 3D brain to help them in surgeries; (v) airports, where the ground crew wears AR glasses to help aircrafts in smooth navigation as well as in landing and take-offs; (vi) historical sites, making AR projections on historical sites, bringing them to life and make the tourists re-live the past era; and (vii) robot path planning; and (viii) IKEA, a furniture company uses an AR application called as IKEA Place which makes the customers see how the furniture and other household items would fit in their houses or offices. Figure 1.14 will help the readers understand the concept of AR in a better way.

Another variation of VR is mixed reality, where the virtual environment is combined with the real world [47]; hence in mixed reality, the user has an interaction with both the real world and virtual world. Although similar, AR and mixed reality are different from each other. Mixed reality is an extension of AR. It brings the best of both real and virtual worlds.

1.1.10 Digital Twin

A digital twin is a replica or virtual representation of physical assets or products. The term digital twin was introduced by Dr. Michael Grieves

FIGURE 1.14 (a) AR in *Pokémon Go* Game. (With permission from Marc Bruxelle/ Shutterstock.com.) (b) AR being used in furniture app. (From https://techcrunch. com/.) [See Ref. 48]. (c) AR being used in phones for suggestions and games. (From https://www.diva-portal.org/.) [See Ref. 49]. (d) AR used by GMaps. (From https:// www.techannouncement.in/.)

in 2002. Before introducing a digital product in the market, its digital twin is introduced to study all the dynamics of the product (from an electron to the whole device). It helps in studying a product's age, product's reaction in different environments and different types of weather conditions (they work with the help of IoT sensors). Feedback from the digital twin of a product influences how the design is built; operation of a device is created to derive the correct output for the product you are making [50,51]. Some examples for digital twins in real-life practices are listed: (i) monitoring wind forms using the digital twin: previously, the

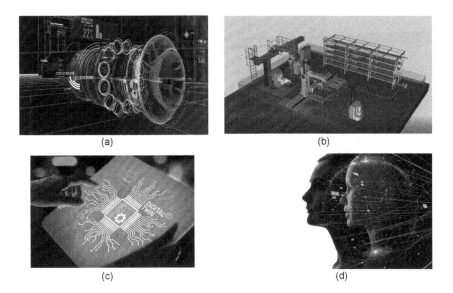

FIGURE 1.15 (a) Digital twin in manufacturing, (b) Digital twin in 3D model of welding production line , (c) Digitial Twin in Nano Technology, and (d) Digitial Twin in Real Time View of physical assets.

companies that produced energy using wind forms were purely based on weather conditions, but the unpredictability of weather, wind direction, and other factors make it a tough task. But using digital twins, now they are able to create fully functioning representation (replica) of wind forms and analyze which wind angle affects the turbine wings the most. By gathering the digital data, they can predict how much energy can be produced by simulating the event in the future using digital twins; and (ii) Digital twin in aerospace: it is difficult for air force to predict the average lifespan of a jet engine; how long can an aircraft use it before the potential risk gets higher. By creating digital twins, we can monitor the engine health and progress through the digital data we have. Figure 1.15 shows some of the digital twin examples.

The basic working of digital twin can be described in a few steps which are mentioned below [52]:

Step 1: Collection of real-time data from installed sensors (mostly IoT sensors).

Step 2: Locally decentralize or centrally store the collected data.

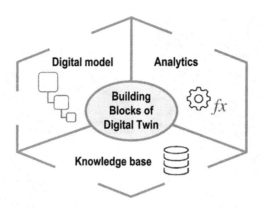

FIGURE 1.16   Building blocks of digital twin.

Step 3: Evaluate and simulate the collected data (manufactured to imitate some other material) in virtual copy of asset.

Step 4: The simulated virtual copy provides information; then, the parameters are applied to real asset.

This integration in real and virtual representation helps in optimizing the performance of real assers.

Digital models, knowledge base, and analytics are the basic building blocks for achieving digital twins (Figure 1.16).

The key benefits of digital twins include (i) improved productivity, (iii) increased reliability and accuracy of product output (iii) risk management, by predicting it through real-time data, and (iv) improved product quality and enhanced product insights. According to PR Newswire, March 2020, DUBLIN [53], (i) digital twin technology is the fourth industrial revolution for the development and marketing of products, (ii) 89% of the IoT platform will use digital twins by 2025, (iii) digital twin will become a standard feature for IoT application until 2027, and (iv) most of the automobile companies understand the importance of digital twins and by the start of 2028, digital twins will become an important element of the automobile industry.

1.2 CONCLUSIONS

As evident, today's world has become highly connected and global with the assistance of the existing and emerging technologies. The speed with which a new technology is coming to the market is amazing, and we as a user and researcher need to keep up with the upcoming technologies.

These technologies have expanded its reach in every sector of the market. This chapter has included major technologies which are changing user's lives at a fast pace. Yet, there are many technologies and interdisciplinary medium which have not been discussed in detail. Some of the technologies (apart from the ones which have been discussed in detail) are worth mentioning here, such as blockchain, business networking fabric, distributed ledger technology, 3D printing, cryptocurrencies, mesh networks, next-generation batteries, and many more. All these technologies aim to make our lives better and easy. Technologies have become an integral part of user's lives connecting everyone to everything in today's time. Although they come with their limitations, they are still better compared to the traditional ways. These limitations are the areas of research and researchers and scientists are coming up with brilliant solutions for them. One of the major limitations of almost every technology is security, and hence security is the major area of research these days, especially in the field of IoT, where minimum amount of work has been carried out in the field of security. Our discussion in this chapter is majorly focused on delivering an introduction on current major technologies and their brief working. The discussion also presents advantages and major applications of all the technologies included in this chapter. Also, this chapter describes the challenges faced while implementing these technologies on real-life applications.

REFERENCES

1. Minsky, M. (1961). Steps toward artificial intelligence. *Proceedings of the IRE, 49*(1), 8–30.
2. Mitchell, T. (1997). Introduction to machine learning. *Machine Learning, 7,* 2–5.
3. Smola, A., & Vishwanathan, S. V. N. (2008). Introduction to machine learning. *Cambridge University, UK, 32*(34), 2008.
4. Prince, S. J. (2012). *Computer Vision: Models, Learning, and Inference.* New York: Cambridge University Press.
5. Jarvis, R. A. (1983). A perspective on range finding techniques for computer vision. *IEEE Transactions on Pattern Analysis and Machine Intelligence, 5*(2), 122–139.
6. Jähne, B., & Haußecker, H. (2000). *Computer Vision and Applications.* San Diego, CA: Elsevier Inc.
7. Russell, S., & Norvig, P. (2002). *Artificial Intelligence: A Modern Approach.* Hoboken, NJ: Prentice Hall

8. Newman, G., Wiggins, A., Crall, A., Graham, E., Newman, S., & Crowston, K. (2012). The future of citizen science: Emerging technologies and shifting paradigms. *Frontiers in Ecology and the Environment, 10*(6), 298–304.

9. Khan, M. A., & Salah, K. (2018). IoT security: Review, blockchain solutions, and open challenges. *Future Generation Computer Systems, 82*, 395–411.

10. Biswas, A. R., & Giaffreda, R. (2014, March). IoT and cloud convergence: Opportunities and challenges. In *2014 IEEE World Forum on Internet of Things (WF-IoT)* (pp. 375–376). IEEE, Seoul, Korea (South).

11. Lee, I., & Lee, K. (2015). The Internet of Things (IoT): Applications, investments, and challenges for enterprises. *Business Horizons, 58*(4), 431–440.

12. Ziegler, J. F. (Ed.). (1992). *Handbook of Ion Implantation Technology* (pp. 92–93). Amsterdam: North-Holland.

13. Gubbi, J., Buyya, R., Marusic, S., & Palaniswami, M. (2013). Internet of Things (IoT): A vision, architectural elements, and future directions. *Future Generation Computer Systems, 29*(7), 1645–1660.

14. Bernstein, E., & Vazirani, U. (1997). Quantum complexity theory. *SIAM Journal on Computing, 26*(5), 1411–1473.

15. Shor, P. W. (1999). Polynomial-time algorithms for prime factorization and discrete logarithms on a quantum computer. *SIAM Review, 41*(2), 303–332.

16. Manin, Y. I. (1980). *Vychislimoe i nevychislimoe (Computable and Noncomputable)*, Moscow: Sov.

17. Feynman, R. P. (1982). Simulating physics with computers. *International Journal of Theoretical Physics, 21*(6/7), 133–154.

18. Feynman, R. P. (1985). Quantum mechanical computers. *Optics News, 11*(2), 11–20.

19. Benioff, P. (1982). Quantum mechanical Hamiltonian models of Turing machines. *Journal of Statistical Physics, 29*(3), 515–546.

20. Deutsch, D. (1991). Rapid solution of problems by quantum computation. *Proceedings of the Royal Society A, 435*, 563–574.

21. Simon, D. R. (1997). On the power of quantum computation. *SIAM Journal on Computing, 26*(5), 1474–1483.

22. Vandersypen, L. M., Steffen, M., Breyta, G., Yannoni, C. S., Cleve, R., & Chuang, I. L. (2000). Experimental realization of an order-finding algorithm with an NMR quantum computer. *Physical Review Letters, 85*(25), 5452.

23. Steffen, M., DiVincenzo, D. P., Chow, J. M., Theis, T. N., & Ketchen, M. B. (2011). Quantum computing: An IBM perspective. *IBM Journal of Research and Development, 55*(5), 13–1.

24. Zhou, Z., Chen, X., Li, E., Zeng, L., Luo, K., & Zhang, J. (2019). Edge intelligence: Paving the last mile of artificial intelligence with edge computing. *Proceedings of the IEEE, 107*(8), 1738–1762.

25. Satyanarayanan, M. (2017). The emergence of edge computing. *Computer, 50*(1), 30–39.

26. Ahmed, E., Ahmed, A., Yaqoob, I., Shuja, J., Gani, A., Imran, M., & Shoaib, M. (2017). Bringing computation closer toward the user network: Is edge computing the solution. *IEEE Communications Magazine, 55*(11), 138–144.

27. Vaquero, L. M., & Rodero-Merino, L. (2014). Finding your way in the fog: Towards a comprehensive definition of fog computing. *ACM SIGCOMM Computer Communication Review, 44*(5), 27–32.

28. Bonomi, F., Milito, R., Zhu, J., & Addepalli, S. (2012, August). Fog computing and its role in the internet of things. In *Proceedings of the first edition of the MCC workshop on Mobile cloud computing* (pp. 13–16), Helsinki, Finland.

29. Yi, S., Hao, Z., Qin, Z., & Li, Q. (2015, November). Fog computing: Platform and applications. In *2015 Third IEEE Workshop on Hot Topics in Web Systems and Technologies (HotWeb)* (pp. 73–78). IEEE, Washington, DC.

30. Yi, S., Li, C., & Li, Q. (2015, June). A survey of fog computing: Concepts, applications and issues. In *Proceedings of the 2015 workshop on mobile big data* (pp. 37–42).

31. Gupta, B. B., & Agrawal, D. P. (Eds.). (2019). *Handbook of Research on Cloud Computing and Big Data Applications in IoT*. IGI Global.

32. Baldini, I., Castro, P., Chang, K., Cheng, P., Fink, S., Ishakian, V., ... Suter, P. (2017). Serverless computing: Current trends and open problems. In Sanjay Chaudhary, Gaurav Somani, Rajkumar Buyya (eds.), *Research Advances in Cloud Computing* (pp. 1–20). Springer, Singapore.

33. McGrath, G., & Brenner, P. R. (2017, June). Serverless computing: Design, implementation, and performance. In *2017 IEEE 37th International Conference on Distributed Computing Systems Workshops (ICDCSW)* (pp. 405–410). IEEE, Atlanta, GA.

34. Jonas, E., Schleier-Smith, J., Sreekanti, V., Tsai, C. C., Khandelwal, A., Pu, Q., ... Gonzalez, J. E. (2019). Cloud programming simplified: A Berkeley view on serverless computing. arXiv preprint arXiv:1902.03383.

35. Li, H., Ota, K., & Dong, M. (2018). Learning IoT in edge: Deep learning for the Internet of Things with edge computing. *IEEE Network, 32*(1), 96–101.

36. Abdi, H. (1994). A neural network primer. *Journal of Biological Systems, 2*(3), 247–281.

37. Sobot, R. (2018). Implantable technology: History, controversies, and social implications [commentary]. *IEEE Technology and Society Magazine, 37*(4), 35–45.

38. Ryan, M. L. (2001). Narrative as virtual reality. *Immersion and Interactivity in Literature*, 357–359.

39. Burdea, G. C., & Coiffet, P. (2003). *Virtual Reality Technology*. John Wiley & Sons.

40. Zheng, J. M., Chan, K. W., & Gibson, I. (1998). Virtual reality. *IEEE Potentials, 17*(2), 20–23.

41. Riener, R., & Harders, M. (2012). Virtual *Reality* in *Medicine*. Springer Science & Business Media.

42. Banerjee, P. P. (2009). Virtual reality and automation. In Springer *Handbook of Automation* (pp. 269–278). Berlin, Heidelberg: Springer.

43. Gupta, J. (2018). "How to transform classroom learning with virtual reality in education", October 12, 2018. [Online]. Available: https://elearningindustry.com/transform-classroom-learning-virtual-reality-education

44. Azuma, R. T. (1997). A survey of augmented reality. *Presence: Teleoperators & Virtual Environments, 6*(4), 355–385.

45. Furht, B. (Ed.). (2011). *Handbook of Augmented Reality*. New York: Springer Science & Business Media.

46. Carmigniani, J., Furht, B., Anisetti, M., Ceravolo, P., Damiani, E., & Ivkovic, M. (2011). Augmented reality technologies, systems and applications. *Multimedia Tools and Applications, 51*(1), 341–377.

47. Cohen, R. (2009). U.S. Patent No. 7,564,469. Washington, DC: U.S. Patent and Trademark Office.

48. Perez, S. (2018). "Wayfair's Android app now lets you shop for furniture using augmented reality", March 20, 2018. [Online]. Available: https://techcrunch.com/2018/03/20/wayfairs-android-app-now-lets-you-shop-for-furniture-using-augmented-reality/

49. Henrysson, A. (2007). Bringing augmented reality to mobile phones (Doctoral dissertation, ACM).

50. Tao, F., Zhang, M., Cheng, J., & Qi, Q. (2017). Digital twin workshop: A new paradigm for future workshop. *Computer Integrated Manufacturing Systems, 23*(1), 1–9.

51. Zheng, Y., Yang, S., & Cheng, H. (2019). An application framework of digital twin and its case study. *Journal of Ambient Intelligence and Humanized Computing, 10*(3), 1141–1153.

52. Haag, S., & Anderl, R. (2018). Digital twin–Proof of concept. *Manufacturing Letters, 15*, 64–66.

53. Cision, "The Future of the Digital Twins Industry to 2025 in Manufacturing, Smart Cities, Automotive, Healthcare and Transport", March 24, 2020. [Online]. Available: https://www.prnewswire.com/news-releases/the-future-of-the-digital-twins-industry-to-2025-in-manufacturing-smart-cities-automotive-healthcare-and-transport-301028858.html.

CHAPTER 2

Artificial Intelligence Innovations

Infrastructure and Application for Advances in Computational Supremacy

Shruti Gupta

National Institute of Technology

Ashish Kumar

Jaypee Institute of Information Technology

Pramod Kumar

Krishna Engineering College

Pastor Arguelles

University of Perpetual Help System DALTA

CONTENTS

DOI: 10.1201/9781003121466-2

2.1 INTRODUCTION

Artificial intelligence (AI) is affecting our lives to a huge extent. Organizations and firms are also taking measures in the direction of acclimatization of AI technology which can provide novel means to execute the tasks along with understanding and recognizing the data patterns in order to achieve utmost productivity. The term AI was coined in 1956 (Chouard and Venema 2015) and is a science based on training the computers in a manner to inculcate human capabilities and intelligence so that they can simulate human behavior. The actions and activities such as learning, perception, inventiveness, interpretation and logic which were limited to humans have now been simulated by technology and being practiced in distinct firms and industries. Soon after AI, the term machine learning (considered as a sub branch or application of AI) was coined in 1959, which enables learning from past cases without being explicitly programmed.

The goal of AI is to build intellectual systems which can execute complex tasks, whereas machine learning aims to solve the specific tasks for which it has been trained. AI and machine learning were going hand in hand until the late 1970s, but with the commencement of the 1980s, the research area of AI shifted to logical and knowledge-based methods more willingly, than algorithms. In the mid-2000s, deep learning came into existence, although its full establishment period was reached by 2010. Deep learning is a subset of machine learning (Figure 2.1) which basically employs neural networks to imitate human brain characteristics or behavior. It is comparatively more powerful than machine learning and can handle large data sets.

AI is utilized in most of the fields as this assists in lessening the risk and also enhances the possibility of attaining accuracy with a higher level of precision. For instance, in the area of healthcare, AI has been a boon by implementing methodologies helpful in drug development, disease diagnosis, patient observation and treatment. AI has spread its wings in the field of education also through personalized learning, monitoring students' progress using virtual mentors. In addition, AI has also contributed to open source robotics and employment of AI-based natural language processing by robots to interact with and provide service to the customers. Another sector which has been advanced by AI is transportation. Innovations of autonomous vehicles have reduced the probability of

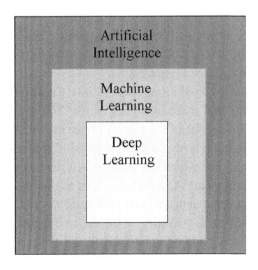

FIGURE 2.1   Subsets of artificial intelligence.

accidents. AI has provided measures for traffic management and pedestrian safety. Beside these, the renewable energy system is another section that has benefited from AI. AI has provided solutions in prediction of wind power generation, solar power generation and hydropower generation. Furthermore, AI has also enhanced the field of remote sensing with services such as automation of multispectral image analysis, advancement in database interrogation and intelligent onboard processing. Moreover, a great transformation has been brought about by AI in the area of finance. AI engines have boosted market analysis, data mining and algorithmic trading. The agriculture sector has also been improved by utilizing AI technology. AI imparts more competent ways to generate, harvest and trade vital crops. Several other applications of AI are in the areas of security, entertainment, service, smart cities, military, e-commerce, media and many more. This chapter discusses all the above mentioned applications and how AI has taken over almost every field and proved to be a great advancement in computational supremacy.

2.2 ORIGINATION AND BACKGROUND OF ARTIFICIAL INTELLIGENCE

Today, AI has provided automation in several fields like assistance in business decision making, productivity enhancement, disease detection, weather forecasting and others. It is also significantly affecting various companies, industries and employment within them. All this leads to knowing about the commencement and getting a view about the evolution of AI. Hence, in this section, a brief and rapid description about the origination and the sets of main events which form the basis of history of AI has been presented (see Figure 2.2).

The seeding of AI started around the 16th–17th century. It began with the invention of the first mechanical calculator (Touretzky 2015) in 1642, followed by the innovation of the binary system in 1679 which gave rise to modern computing (Hardy 2020). In 1927, a film based on robots was released, which marks the debut of AI. After this the theory started to take its practical formation in 1943, with the birth of the first neural network model (McCulloch-Pitts neurons) (Warren and Pitts 1943). A mathematical algorithm based on the working of human neurons was proposed by a neurophysiologist and a mathematician. The model was also implemented with electric circuits, which intended to imitate human brain procedure. In 1950, a successful Turing test (Saygin et al 2000) was performed by a

FIGURE 2.2   History of artificial intelligence.

British mathematician. In the test the machine had to participate in a conversation with a human via writing and convince him that the machine itself was also human. The motive of the test was to discover real intelligence in a computer. In 1952, a checkers program for an IBM computer was developed by Arthur Samuel. The program was based on the game of checkers (John and Ea 1990). It was a type of automatic learning based on the fact that the more number of times the game is played, the better the system will become. Finally the term "Artificial Intelligence" was introduced in a summer conference by John McCarthy in 1956 at the Dartmouth College. In 1957, Rosenblatt set the foundation of deep neural networks. He designed the neural network known as "Perceptron." The aim of "Perceptron" (Widrow and Lehr 1990) was to identify patterns, shapes and images. In 1966 the first chatbot ELIZA was created for solving mathematical problems. The era between 1974 and 1980 was known as the first AI winter as during this phase there was scarcity of funding for AI research, although in 1979, "Stanford Cart" (Moravec 1983), a remotely controlled TV-equipped mobile robot, was developed to navigate obstructions in a room by itself. After the second AI winter, a boom happened with Deep Blue, an IBM computer, which beat the world chess champion in 1997 (Hardy 2020). In 2006, the term "deep learning" was coined that allows the algorithms to observe and differentiate objects and texts in videos and images. ImageNet (2009), a large scale hierarchical database; IBM's Watson (2011), a natural language based problem solving computer, won Jeopardy and defeated two earlier champions; Google Brain (2012), a deep neural network for pattern recognition in images and videos; DeepMind (2014), a company acquired by Google, used to play human level video games; OpenAI (2015); Lipnet (2016), Stopage Online trolling (2017), Google Duplex (2018), an assistant to engage appointments over the phone; DeepSpeed and OpenAI's GPT-3 (2020) were some of the constituents which have taken AI to a remarkable level (Hardy 2020), (MLK 2019), (Foote 2019). History of AI predicts that the future lies in it through several ways.

2.3 DIFFERENT TYPES OF AI

AI could be categorized into several ways (Wang and Siau 2019). However, there are two broader aspects of classifying AI which are based on their capabilities and functionality. On the basis of capabilities, AI can be classified into three types: Narrow (Weak) AI, General AI and Super AI.

Narrow AI is the simplest form of AI which has the capability to perform pre defined tasks and in a pre defined scenario. The machine in this stage does not have its own thinking ability and performs only in a limited domain for which it has been trained. Narrow AI can be unsuccessful in unpredictable ways if it goes beyond its limits. Popular examples of Narrow AI are Apple Siri, Alexa by Amazon, IBM's Watson supercomputer, self-driving cars, Alpha-Go, Sophia the humanoid, playing chess, buying recommendations on e-commerce sites, speech and image recognition, etc.

General AI is the intelligence which has the ability to carry out logical tasks with efficiency equivalent to that of a human. General AI aims to develop a system that is clever and intelligent and should possess the thinking, understanding and acting ability similar to that of a human. Presently, all the existing intelligent systems are classified under the category of Narrow AI and no system lies under General AI. The development of General AI systems requires efforts and is a prolonged process; hence it is still under research. Super AI is a stage of intelligence at which machines may perhaps exceed human intelligence, and be able to execute every task in an enhanced manner in comparison to humans. Some of the significant features of Super AI include thinking ability; reasoning ability; puzzle solving skill; capability to take decisions, learning and planning. Super AI is still a speculative conception and its implementation would be a drastic world changing job. Other than this, on the foundation of functionality, there are four types of AI: Reactive machines, Limited memory, Theory of mind and Self-awareness. Reactive machines are the simplest type of AI which perform fundamental operations. Such an AI doesn't have any past conceptions or experiences, nor any memory store of what has occurred previously. These machines just concentrate on current situations and respond to them in the best possible manner. Examples of Reactive AI are IBM's Deep Blue system and Google's AlphaGo.

Limited memory AI can accumulate previous experiences or some data for a short duration of time and based on that can make knowledgeable and better judgments for future actions by analyzing the past data from its storage. One of the significant examples of Limited memory AI is self-driving cars. These cars can save the latest speed of neighboring cars, react to road signs and traffic lights and make better driving judgments. Theory of mind AI is based on the machine understanding of human emotions, ideas, values; and the ability of social interaction is just similar to that of humans. It focuses on awareness about the actions of other objects and creatures and then reacts accordingly. There is still no development of this type of AI machines, but research work is still going on for constructing such AI machines. Self-awareness AI is the future outlook of AI and would be cleverer in comparison to humans. These machines will be highly intelligent, and will possess their own awareness, affection, and self-responsiveness. Self-awareness AI is a speculative theory and is not present in reality.

2.4 CHALLENGES IN AI

Although AI possesses enormous potential and the ability to produce intelligent machines, it does have its own sets of challenges (Perc et al 2019). Some of the significant challenges in AI have been addressed in this section. One of the important challenges is the requirement of high computing power for implementing complex AI algorithms and techniques. Arranging funds for obtaining supercomputer is not an easy task, especially for startups and small businesses. Another challenge is the belief of people and laymen worldwide on AI. People are not familiar with deep learning models and their functioning, and are therefore not able to grasp its concept and usage. Lack of technical knowledge in entrepreneurs for establishing AI applications in the enterprise is a roadblock in AI advancement. Proficient human resources could help in adopting AI solutions. Human level expectation from AI is another issue. With the deployment of AI, companies are boasting very high accuracy, but the fact is that AI still cannot beat the human level caliber. For implementing AI, collection and utilization of appropriate data sets is essential and maintaining its privacy and security is another significant challenge as it could be utilized for inadequate activities. The ethical challenge in AI is the biasness present in the algorithms. The algorithm performance is based on the data using which it has been trained. If the training data is bad i.e., fortified with race, gender, community, religion etc., then it will result in unfair and unethical outcomes. A company may face legal issues also if the AI application consists of faulty data governance and algorithms. The above discussed challenges appear quite discouraging and disparaging for humanity, but with the combined endeavor of people, these challenges could be resolved very efficiently. As said by Microsoft, the future generation of engineers needs to make themselves proficient in these revolutionary and new-fangled technologies. UpGrad has been offering programs also on these technologies and several of their students are working in Google, Amazon, Microsoft, Amazon, Visa and various other affluent companies.

2.5 ARTIFICIAL INTELLIGENCE: INFRASTRUCTURE AND FRAMEWORK

2.5.1 AI Agents and Environment

AI is the course of agents and their environment. Agents could be described as an entity that can observe their environment utilizing sensors and take actions on that environment using effectors. Agents are described with respect to their rationality. Rationality refers to the agent's performance evaluation criterion, its past awareness about its environment, best feasible actions that an agent can carry out and the series of percepts. On the basis of their level of perceived intelligence and potential, agents can be divided into four groups:

- Simple reflex agent (take actions on the basis of the current percept)

- Model-based reflex agent (take actions on the basis of the existing model),

- Goal-based agents (take actions in order to achieve goal),

- Utility-based agent preference (take actions on the basis of utility for each state) and

- Learning agent (take actions on the basis of past experiences).

The surroundings where the agent lives, operates, senses and acts is known as environment. From the agent's perspective, the environment has different characteristics.

- On the basis of the scope of observation in an environment, an environment could be fully *observable, partially observable* or *unobservable.*

- The environment is *deterministic* if its next state could be determined on the basis of a specific state, otherwise environment is *stochastic.*

- *Discrete* environments are those on which a finite set of percepts can lead to final result of the task, e.g., a chess game, while *continuous* environment depends on indefinite and rapidly varying data sources, e.g., self-driving car or drones.

- The environment which keeps on altering itself with respect to time is called *dynamic*, else *static*.

- An environment involving one agent is called a *single agent* environment while an environment having more than one agent is called a *multi-agent* environment.

- An environment is said to be *competitive* if the agents compete with each other to obtain optimal outcome; on the other hand if the agents cooperate among themselves to obtain a specific outcome it is known as a *collaborative* environment.

- If an agent could acquire full and precise information regarding the state's environment, then the environment is said to be an *accessible* environment otherwise it is known as *inaccessible*.

2.5.2 AI Search Algorithms

Another significant area in AI is search algorithms. In order to solve any specific problem or attain any goal, agents perform some sort of search algorithms. Search algorithms could be described or compared on the basis of certain features. These features are completeness (if it offers assurance to return a solution if it exists for any arbitrary input), Optimality (assurance to return an optimal solution in case solution is present), time complexity (quantification of time taken by an algorithm to finish the task) and space complexity (measure of storage space needed during the search). With respect to search problems, search algorithms could be classified into uninformed search algorithms and informed search algorithms (Alhassan et al 2019). Informed search algorithms possess information on the goal state acquired using a function which evaluates the closeness between a given state and the goal state. This information assists in proficient searching. Some of the informed search techniques are best-first search, Greedy search and A* search. Uninformed search algorithms possess no extra information on the goal node except the one made available in the problem description. The strategies to attain the goal state from the initial state vary merely by the sequence and length of actions. Uninformed search techniques are breadth-first search, depth-first search, uniform cost search, bidirectional search, depth limited search, etc.

2.5.3 Tools and Techniques Practiced in AI

AI is a multidisciplinary area which has grabbed researchers from distinct fields. Some of the significant AI tools (Sara Devis and Bekar 2017) are Scikit Learn, TensorFlow, Theano, Caffe, MxNet, Keras, PyTorch, CNTK,

Auto ML, OpenNN, H20: Open Source AI Platform and Google ML Kit. Although there is a wide range of AI techniques, a few of the fundamental techniques are grouped and discussed in the following sections.

2.5.3.1 Machine Learning Techniques

Machine Learning techniques recognize patterns and associations from data to attain an optimal result. These are commonly divided into three categories: supervised learning, unsupervised learning and reinforcement learning.

2.5.3.1.1 Supervised Learning   Supervised learning (Mohri et al 2018) is one of the significant sections of machine learning these days. It is the initial point of expedition for most of the machine learning professionals. Supervised techniques accustom the model based on a known training set and then use the model for predicting the output of unforeseen inputs. The motive of the machine learning model here is to determine a mapping function from a training set of input-output pairs, which can identify the outcomes for the input data set other than training data. The general form of supervised learning can be depicted as:

$$n = f(m)$$

where, "n" is the output data, "m" is the input data and "f" is the mapping function. The asset of supervised machine learning is: it is quite fast and authentic. It also allows utilizing the data to recognize and avert undesirable results or enhance the considerable results for their target objects. Basically, supervised learning techniques have been divided into two types: regression techniques and classification techniques.

2.5.3.1.1.1 Regression Techniques   Regression techniques are based on the formation of a mathematical model by conducting predictive analytics among the input (independent) and output (dependent) variables, using the labeled training data set. Here, the model produces a continuous value output with minimal error. Common regression algorithms include linear regression, logistic regression and polynomial regression.

2.5.3.1.1.2 Classification Techniques   Classification is a type of learning which produces discrete outcomes in the form of categories or classes. The classification algorithm maps the input data into distinct classes on the basis of training data set. The classification is known as binary classification if the prediction is among the two classes. If more than two classes are concerned then it is known as multiclass classification. There are many classification algorithms present, some of the prominent ones are:

Support Vector Machine, decision tree, neural network, Random Forest Classifier and Naive-Bayes classifier.

2.5.3.1.2 Unsupervised Learning   Unsupervised (Duda et al 2008) machine learning algorithm has quite opposite characteristics in comparison to supervised machine learning algorithm. In this type of learning no pre-classified data or training data set has been provided; instead the model itself has to draw inferences, find hidden patterns and discover structures from the input data sets without any supervision. Unsupervised learning techniques are usually used for pre-processing the data, before applying any further rigorous computations. This type of machine learning is quite complex, less precise and less reliable in contrast to supervised learning. Unsupervised learning comprises distinct clustering methods, dimensionality reduction techniques and associative analysis.

2.5.3.1.2.1 *Clustering Methods*   Clustering (Rokach and Maimon 2005) is the dissection of a data set into distinct subsets or clusters on the basis of some principle or fact, such that the data points in the same cluster show analogous behavior or properties in comparison to the data points in other clusters. There are distinct clustering techniques (K-means, Hierarchical Clustering, Density based Clustering, etc.) present which formulate different postulations on the composition of the data set, in the form of some similarity or dissimilarity metric among the constituents of the same or different clusters.

2.5.3.1.2.2 *Dimensionality Reduction Techniques*   Dimensionality reduction is one of the beneficial practices applied to high dimensional data, so as to decrease the number of dimensions (features). Low dimensional data is probably easy to analyze in comparison to high dimensional data and does not endure from the curse of dimensionality. Dimensionality reduction is mainly performed in two ways: either selecting the significant features from the feature set or by constructing a new feature space from the original features. Some of the commonly used dimensionality reduction techniques are Principle Component Analysis (PCA), Independent Component Analysis (ICA), etc.

2.5.3.1.2.3 Associative Analysis Association analysis (Jiawei et al 2012) is a methodology of identifying hidden patterns and relationships in a large data set. The identified relationships could be characterized in the guise of association rules. Association rule learning aims to determine powerful rules explored in databases on the basis of some means of interest. This concept is often practiced for market basket analysis. The potency of an association rule depends on two measures, namely, support and confidence. Support identifies how frequently a rule is pertinent to a given data set while confidence measures the consistency of the presumption made by the rule. Distinct methods for generating association rules are Apriori Algorithm, Frequent Pattern Growth (FP Growth), etc.

2.5.3.1.3 Reinforcement Learning   Reinforcement learning (Kaelbling et al 1996; Lucian et al 2010) is a type of machine learning in which the model is made to learn on the basis of its interaction with the environment. The algorithm works on the basis of trial and error. If the model produces encouraging outcomes, it will be rewarded or reinforced otherwise it will be punished. Hence, the goal of the model is to maximize the reward and keep on iterating until it achieves the goal. Reinforcement learning is yielding remarkable benefits in the generation of power.

2.5.3.2 NLP (Natural Language Processing)

Basically, NLP (Young et al 2017) is a method of communication between computers (intelligent systems) and human beings. Computers utilize this technique to understand, interpret and manipulate human language. There are two constituents of NLP, namely, Natural Language Understanding (NLU) and Natural Language Generation (NLG). NLU consists of mainly two steps: first the input in the form of natural language is mapped into convenient representations. Second, the analysis of distinct facets of the language is performed. NLG is the process of transforming the internal representation into some relevant phrases and sentences of a natural language. NLG is composed of the following tasks: text planning, sentence planning and text realization. Some of the applications of NLP can be seen in Google assistant, call centers in the form of IVR (Interactive Voice 12qResponse), Apple Siri, Microsoft Cortana, and Alexa from Amazon.

2.5.3.3 Fuzzy Logic

Fuzzy Logic (FL) (Bělohlávek et al 2017) is a technique of reasoning that is similar to that of human reasoning. In our everyday life situations may arise where we are not able to decide or identify between true or false. Fuzzy logic provides us the intermediary likelihood between true or false. Normally, fuzzy logic systems are utilized for commercial and practical purposes together, e.g., for controlling machines and consumer products, for acceptable reasoning and deals with uncertainty in engineering. The architecture of FL is composed of Rule Base (contains a set of rules for decision making), Fuzzification (transform inputs in the form of crisp numbers into fuzzy sets), Inference Engine (decides the rules which are to be fired based on the input) and defuzzification (convert the fuzzy sets into a crisp value). Applications of fuzzy logic could be found in aerospace field for altitude control, in automotive systems for controlling speed and traffic, in decision making support systems and expert systems.

2.5.3.4 Nature Inspired Intelligence Techniques

Natural and biological systems were proved as a significant resource from where scientists derive motivation to develop innovative computational methods (Antonopoulos et al 2020). In the area of AI, nature-inspired algorithms have been used for searching and planning tasks, i.e., to identify the series of actions required for attaining an agent's goals. Evolutionary algorithms are based on heuristic techniques that utilize biological evolution-inspired approaches by computationally imitating some of their key concepts, such as replication, mutation, recombination, and selection. Evolutionary learning algorithms comprise genetic algorithms (GA), differential evolution (DE), learning classifier systems (LCS), etc. Another set of nature inspired algorithms are swarm intelligence algorithms which are based on the intelligent behavior of biological swarms and how simulating their behaviors' various tasks could be solved. Some of the swarm intelligence algorithms are Particle Swarm Optimization (PSO), Ant Colony Optimization (ACO), Grey Wolf Optimizer, Artificial bee colony, etc. Besides the abovementioned algorithms, other nature-inspired meta-heuristics algorithms are the *Artificial Immune System* (AIS) algorithm based on the biological immune systems procedure, *Simulated Annealing method* based on annealing concept, and Wind Driven Optimisation (WDO) algorithm based on atmospheric motion.

2.6 ROLE OF ARTIFICIAL INTELLIGENCE IN ADVANCEMENT AND EMERGENCE OF DISTINCT FIELDS

2.6.1 AI in Healthcare

In the area of healthcare, AI has been a boon by implementing methodologies (Jiang et al 2017) helpful in drug development, assisting physicians to take enhanced clinical decisions, disease diagnosis, patient observation and treatment. Basically, the role of AI in the area of healthcare could be broadly categorized into vital biomedical research, translational medical research and clinical practice (see Figure 2.3).

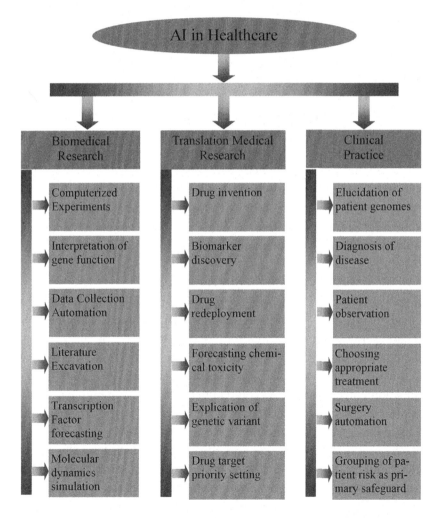

FIGURE 2.3   Applications of AI in healthcare.

The impact of AI in biomedical research mainly includes computerized experiments, interpretation of gene function, data collection automation, literature excavation, transcription factor forecasting and molecular dynamics simulation. Contribution of AI in translational medical research comprises drug invention, biomarker discovery, drug-target priority setting, drug redeployment, forecasting of chemical toxicity and explication of genetic variant. In the category of clinical practice AI has shown potential in elucidation of patient genomes, diagnosis of disease, patient observation, choosing appropriate treatment, surgery automation, grouping of patient risk as primary safeguard, etc. AI performance is at the expert level in the assessment of certain radiology images (for example cardiovascular MRI images or X-ray images for distal radius fracture); dermoscopic melanoma diagnosis and fundus photograph evaluation for diabetic retinopathy etc. where AI does the majority of the task, and clinicians only confirm the diagnosis (Yu et al 2018).

Some of the examples of AI in healthcare which prove its potential in this field are as follows: PathAI (medical technology company) is working on AI-based technology to guide pathologists in performing error free cancer diagnosis; Enlitic, an AI-based company aims to build medical tools to rationalize radiology diagnoses; Buoy Health is an AI-based chatbot to analyze patient symptoms and suggest cures based on its diagnosis; Freenome utilizes AI at screenings so as to detect cancer at its initial stage and Beth Israel Deaconess Medical Center is also deploying AI to identify deadly blood diseases at a premature stage.

2.6.1.1 Role of AI in Pandemic of COVID-19 Disease

The epidemic of COVID-19 is increasing worldwide each day. In the global fight against COVID-19, medical imaging such as computed tomography (CT) and X-ray play a significant role (Shi et al 2020). The merging of these imaging tools and technologies with AI has strengthened their power additionally and provided a great assist to medical specialists. Image acquisition entrusted with AI could appreciably assist in automating the scanning process and also change the work format with the least physical interaction with patients and hence bestowing the image technicians with the best safeguard. AI has also advanced work efficiency by precise demarcation of infections in X-ray and CT images, simplifying consequent assessment. Furthermore, radiologists have obtained a great assist from computer-aided platforms in taking significant clinical decisions for disease identification, monitoring and forecast.

2.6.2 AI in Education

AI has spread its wings in the field of education (Chassignol et al 2018) through personalized learning, monitoring students' progress using virtual mentors, providing smart content which includes vast learning matter from digitized textbooks to custom-made interfaces and much more. AI tools and devices have provided the basis for global learning by accomplishing global classrooms available to all regardless of their language or disabilities. AI has offered modes for interaction with knowledge with the use of voice assistants such as Amazon Alexa and Google Home. AI has also provided ways for global learning and innovation of several platforms and applications for education. There are many AI-based platforms present such as Third Space Learning, Little Dragon, CTI, Brainly, Carnegie Learning and ThinkerMath which can evaluate the degree of knowledge, provide rearward communication, offer a plan for progress, etc.

2.6.3 AI in Robotics

Robotics is another area assisted by AI. Robotics is an environment in which the mechanical human beings (known as Robots) are developed and built. For automating tasks interior and exterior to the industrial unit, AI and robotics are a strong combination (Bogue 2014). AI is gradually becoming a regular existence in robotic solutions, incorporating versatility and erudition abilities in formerly rigorous applications. AI also acts as a tool in applications related to robotic assembly and robotic packaging by providing faster, cheaper and more precise packaging. In addition, AI has also contributed to open source robotics and employment of AI-based natural language processing by robots to interact with and provide service to the customers. "Sophia," the human-like robot developed by the Hong Kong company named Hanson Robotics, is a great example of the emergence of AI in this field.

2.6.4 AI in Transportation

Today, the transportation segment has progressed to a stage where vehicles can maneuver and travel devoid of any human support. Technological expansions have facilitated the evolvement of transportation sector in its expedition of modernization and advancement. AI is one such new-fangled technology that has furnished this sector (Abduljabbar et al 2019) (Boukerche et al 2020). In the transportation sector, major challenges such as capacity issues, protection, reliability, environmental contamination and wasted energy have given liberal prospect for AI innovations. Some of the recent innovations of AI in the transportation industry are autonomous buses and trucks which aim to reduce expenses, lessen emissions, increase road safety, and reduce accident probability in contrast to conventional trucks having human drivers; the American company Local Motors have launched *Olli*—a self-driving, "cognitive" electric shuttle for executing tasks such as carrying passengers to desired locations, giving advice regarding neighboring sights, etc. In the area of railway transportation, *sensors and cameras* are equipped in order to monitor tracks (front and back), which helps in taking onboard real-time decisions; provision of *driverless trains* which can transport larger numbers of passengers. In air transportation the initiation of autonomous planes could provide a massive gain; traditional passports could soon be replaced by face scanners to verify the passengers' identity before permitting them to board. AI has ensured safety of travelers, pedestrians, and drivers by prediction and detection of accidents; reduces traffic congestion, lessens carbon emissions, and also minimizes the overall financial expense. AI has provided measures for traffic management and optimal route scheduling utilizing data analytics in logistics with least wait times.

2.6.5 AI in Renewable Energy

Renewable energy systems have become one of the main sources of interest and concern, due to the unavoidable exhaustion of fossil reserves and the awareness that it is the only feasible way of attaining a clean energy vision. The conventional energy obtained by combustion of fossil fuels such as coal, oil and natural gas in the power sector is resulting in large emissions of carbon-di-oxide, harmful gases and other pollutants. Renewable energy (Lund 2007) is gradually substituting conventional energy owing to its several benefits to the environment.

With the emergence of renewable energy technologies, various implementation problems have also been encountered (see Figure 2.4) as the production of energy relies on the environmental elements which cannot be entirely forecasted, restrained or outlined in advance. For example, in large wind farms, regular condition monitoring of the wind turbine is needed for maintenance (Stetco et al 2019); also, the production of energy is unpredictable as it depends on the wind speed, due to which the designing of the energy grid becomes difficult. Similar is the case with solar energy (Voyant et al 2017) and hydro energy as its production also varies with weather

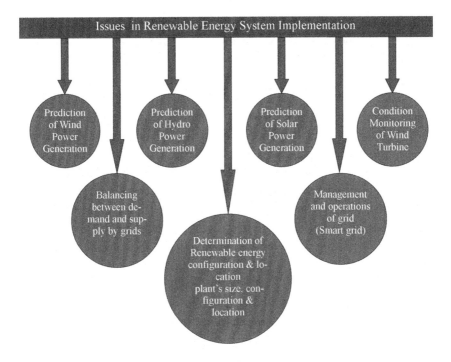

FIGURE 2.4   Issues in renewable energy.

variability. AI has provided solutions in the prediction of wind power generation, solar power generation and hydropower generation. Problems are faced in solar panel installation, to adjust the angle of solar panel with the sun in order to obtain the maximum amount of power per day. Another challenge is the designing of smart energy grid which should be flexible, productive and worthwhile (Perera et al 2014), and keep track of the power produced and demand, without human intervention. Deciding the renewable power plant optimal location, size and configuration is yet another significant application of machine learning techniques in relation with renewable energy. These power plant parameters rely on several factors such as closeness to population region, local climatic variations, topography, organization cost and other provisions. All these problems and many more are proving to be a hindrance in the booming of renewable power. AI has offered ways for condition monitoring of wind turbines; determination of the renewable energy plant's size, configuration and location; balancing between demand and supply of power by grids and management and operations of the grid. AI technology could perhaps prove to be the savior from these problems and could modernize the renewable energy region (Perera et al 2014; Mosavi et al 2019; Sharifzadeh et al 2019).

2.6.6 AI in Geoscience and Remote Sensing

Remote sensing is a tool which helps in the acquisition and analysis of large volume of data in distinct spectral, temporal and spatial provinces. With the enclosure of AI techniques in the satellite image processing and analysis procedure the system has become more advanced and mechanized (Lary et al 2016; Schulz et al 2018). AI has shown successful results in the field of remote sensing with services such as automation of multispectral image analysis, advancement in database interrogation and intelligent onboard processing. Other than this, distinct machine learning techniques have been deployed for extracting the information automatically from data by utilizing the statistical and computational methods. The applications of machine learning techniques in remote sensing and geosciences are not only limited to this but have also expanded its wing in ocean products, tracing gases, vegetation interpretation using indices, portrayal of rock mass, liquefaction process, ground motion parameters, construing the remote sensing image, aerosol products, etc.

2.6.7 AI in Finance

The world of banking and the financial industry has been offered a mode by AI to satisfy the requirements of clients who want better, more opportune, secure methods to access, pay, save and invest their funds (Blooshi and Nobanee 2020). Today, in order to systematize several activities, sustain bookkeeping, invest in stocks, and deal with property, banks employs AI systems. Algorithmic trading utilizes complex AI systems to perform trading decisions at a frequency higher than the capability of any human. In the investment management commerce robo-advisors are fetching attention. Robo-advisors are being deployed for giving monetary guidance and portfolio management with the least human involvement. There are many AI-based companies which are helping the financial industry in the underwriting process; a few of them are DataRobot which assists in constructing predictive models that augment decision-making regarding issues such as fraudulent credit card transactions, blockchain, lending, etc., ZestFinance, which builds an AI-powered underwriting solution namely the Zest Automated Machine Learning (ZAML) platform, aids companies to evaluate borrowers having less or no credit information or records; Scienaptic Systems and underwrite.in are also companies which are working in the same domain.

2.6.8 AI in Agriculture

Agriculture and farming is one of the oldest and most significant professions in the world. With the increase in the world population, there is more scarcity of land and people need to be more creative so as to produce more crops using less land. AI technology in agriculture industry (Chukwu 2019; Smith 2018) is helping farmers to overcome the major challenges and issues they face in agriculture production. Deployment of AI technology is assisting to organize and handle any unwelcome natural conditions. Today, large proportions of start-ups in agriculture are practicing AI-based perspective to amplify the efficiency of yield production. AI imparts more competent ways to generate, harvest and trade vital crops. AI is also utilized in applications dealing with preset machine regulations for weather forecasting and detection of disease or pests. Overall we could say that AI has paved the way toward smart agriculture. Some of the major innovations in agriculture are *John Deere Section Control*, a farmer tool for sowing the seeds with precision by avoiding both overlapping and skipping excessive space when planting crops which also briefs about the location where it cultivates the best; *Robots* are being utilized for harvesting; deployment of *drones* for crop and field monitoring, for example *FarmShots* is a startup that aims to interpret agricultural information extracted from the images captured by the drones and satellites; *See and Spray Model* for controlling pests and weeds, reference to *AI based algorithms* for taking crop production based decisions; operation of *Autonomous tractors* for performing multi-task, AI-powered *Indoor Farming* and employing *Chatbots* for solving farmers' queries.

2.6.9 Other Areas

Several other applications of AI are in the sectors of security, entertainment, service, smart cities, military, e-commerce, media and many more. AI is making much progress in the scientific and research sector also. AI has the ability to handle and analyze large data sets in comparison to human beings, which makes AI deployment quite useful where there is large volume of data. The scope of AI in data analytics is growing swiftly. Helixa. ai is one of the examples of such an AI application. Another area getting advantage from AI is Cybersecurity (Wirkuttis and Hadas Klein 2017).

Organizations and firms keep transmitting their information to IT networks and the cloud, hence there is a significant risk of hackers. Cognitive AI has proved quite useful in this field as it identifies and evaluates threats and also provides perception to the analysts for taking well-informed judgments. For example, the open platform IBM Resilient provides a framework and hub for handling security responses. One more area impacted due to AI is fraud detection. Recurrent Neural Networks have the ability to detect frauds at their initial level. AI also plays a significant part in tactical games (Yannakakis and Togelius 2018) such as chess and poker by suggesting several moves and possible situations on the basis of heuristic knowledge.

2.7 FUTURE SCOPE

AI has an effect on the future of nearly every organization and every human. In the years to come, AI will go on nurturing in a mode to advance the quality of life by facilitating humans to accomplish their tasks more efficiently and quickly. AI has led to a progression in business gain, proficiency and extensive-level economic advancement. In the present era, the rapidity of evolution is quicker than before. Advancement in technology could be seen at such a fast rate, from autonomous vehicles in transportation to voice assistants such as Alexa and SIRI; from alpha zero to the robot Sophia, the list has no bounds. Today's AI is called as weak AI owing to its boundaries. However, the outlook of AI is to create strong AI. Presently, AI has the limitation to perform better than humans in some particular jobs only, although in the future it is projected that AI could defeat humans in all tasks relating to perception, reasoning, understanding, intelligence and awareness. Certainly it has its effects which would be constructive as well as destructive.

2.8 SUMMARY

The motive of this chapter was to focus on the basis of AI, its innovations in distinct fields, and how its applications are advancements in computational preeminence. This chapter initiates with the brief introduction of AI along with its origination and background. Sequentially distinct types of AI have been briefed followed by the challenges faced in AI. Distinct types of AI on the basis of capabilities and functionalities have also been discussed. The infrastructure of AI has also been depicted, which includes AI agents and environment; AI search algorithms followed by tools and techniques employed in the AI system. There are a number of approaches present in AI and the area is continuously expanding. In this chapter, some of the popular approaches have been discussed, including machine learn ing techniques and its types (supervised learning, unsupervised learning and reinforcement learning); natural language processing, fuzzy logic, and evolutionary optimization techniques. Further, the role of AI in advancement and emergence of distinct fields have been analyzed. The fields which were strengthened with AI are Healthcare, Education, Robotics, Transportation, Renewable Energy, Geoscience and Remote Sensing, Finance, Agriculture, gaming, cyber security, etc. This chapter concludes with a overview about the future scope of AI in distinct emerging sectors.

REFERENCES

Abduljabbar, R., Dia, H., Liyanage, S., et al. 2019. Applications of artificial intelligence in transport: An overview. *Sustainability* 11: 189.

Alhassan, T. Q., Omar, S. S. and Elrefaei, L. A. 2019. Game of Bloxorz solving agent using informed and uninformed search strategies. *Procedia Computer Science* 163: 391–399.

Antonopoulos, I., Robu, V., Couraud, B., et al. 2020. Artificial intelligence and machine learning approaches to energy demand-side response: A systematic review. *Renewable and Sustainable Energy Reviews* 130: 109899.

Bělohlávek, R., Dauben, J. W. and Klir, G. J. 2017. *Fuzzy Logic and Mathematics: A Historical Perspective*. Oxford University Press, USA.

Blooshi, L. A. and Nobanee, H. 2020. Applications of artificial intelligence in financial management decisions: A mini-review. *SSRN Electronic Journal*.

Bogue, R. 2014. The role of artificial intelligence in robotics. *Industrial Robot* 41: 119–123.

Boukerche, A., Tao, Y. and Sun, P. 2020. Artificial intelligence-based vehicular traffic flow prediction methods for supporting intelligent transportation systems. *Computer Networks* 182: 107484.

Chassignol, M., Khoroshavin, A., Klimova, A., et al. 2018. Artificial intelligence trends in education: A narrative overview. *Procedia Computer Science* 136: 16–24.

Chouard, T. and Venema, L. 2015. Machine intelligence. *Nature* 521: 435.

Chukwu, N. E. 2019. Applications of artificial intelligence in agriculture: A review. *Engineering, Technology & Applied Science Research* 9: 4377–4383.

Duda, R. O., Hart, P. E. and Stork D. G. 2008. Unsupervised learning and clustering. *In Machine Learning Techniques for Multimedia* (pp. 51–90). Springer, Berlin, Heidelberg.

Foote, K. D. 2019. A Brief History of Machine Learning. Data University. https://www.dataversity.net/a-brief-history-of-machine-learning/ (accessed September 21, 2020).

Hardy, Q. History of Machine Learning. Build with Google cloud. https://cloud. withgoogle.com/build/data-analytics/explore-history-machine-learning/ (accessed September 20, 2020).

Jiang, F., Jiang, Y., Zhi, H., et al. 2017. Artificial intelligence in healthcare: Past, present and future. *Stroke and Vascular Neurology* 2:e000101.

Jiawei, H., Pei, J. and Kamber, M. 2012. *Data Mining: Concepts and Techniques* (Third edition). Morgan Kaufmann Publishers, USA.

John, M and Ea, F. 1990. Arthur Samuel: Pioneer in machine learning. *AI Magazine* 11(3): 10.

Kaelbling, L.P., Littman, M. L. and Moore, A. W. 1996. Reinforcement learning: A survey. *Journal of Artificial Intelligence Research* 4: 237–285.

Lary, D. J., et al. 2016. Machine learning in geosciences and remote sensing. *Geoscience Frontiers* 7: 3–10.

Lucian, B., Babuska, R., Schutter B. D., et al. 2010. *Reinforcement Learning and Dynamic Programming using Function Approximators.* Taylor & Francis CRC Press, Boca Raton, FL.

Lund, H. 2007. Renewable energy strategies for sustainable development. *Energy* 32(6): 912–919.

MLK. 2019. Brief History of Deep Learning from 1943–2019 [Timeline]. MLK: Making AI Simple. https://machinelearningknowledge.ai/brief-history-of-deep-learning/ (accessed September 20, 2020).

Mohri, M., Rostamizadeh, A. and Talwalkar, A. 2018. *Foundations of Machine Learning* (Second edition). The MIT Press, London.

Moravec, H. P. 1983. The Stanford cart and the CMU rover. *Proceedings of the IEEE* 71: 872–884.

Mosavi A., et al. 2019. State of the art of machine learning models in energy systems, a systematic review. *Energies* 12: 1301.

Perc, M., Ozer, M. and Hojnik, J. 2019. Social and juristic challenges of artificial intelligence. *Palgrave Communications* 5: 61.

Perera, K.S., Aung Z. and Woon W.L. 2014. Machine learning techniques for supporting renewable energy generation and integration: A survey. In: Woon, W., Aung, Z., Madnick, S. (eds) *Data Analytics for Renewable Energy Integration*, vol. 8817. Springer, Cham.

Rokach, L. and Maimon, O. 2005. Clustering methods. In: Maimon, O., Rokach, L. (eds) *Data Mining and Knowledge Discovery Handbook.* Springer, Boston, MA.

Sara Devis, S. and Bekar, J. 2017. Top 12 AI Tools, Libraries, and Platforms. Dzone: A Devada Media Property. https://dzone.com/articles/ai-tools-and-libraries (accessed September 20, 2020).

Saygin, A. P., Cicekli, I. and Akman, V. 2000. Turing test: 50 years later. *Minds and Machines* 10: 463–518.

Schulz, K., Hänsch, R. and Sörgel, U. Machine learning methods for remote sensing applications: An overview. In *Proceedings of SPIE 10790, Earth Resources and Environmental Remote Sensing/GIS Applications IX*, Berlin, Germany, 1079002, September 2018.

Sharifzadeh, M., Sikinioti-Lock, A. and Shah, N. 2019. Machine-learning methods for integrated renewable power generation: A comparative study of artificial neural networks, support vector regression, and Gaussian Process Regression. *Renewable and Sustainable Energy Reviews* 108: 513–538.

Shi, F., Wang, J., Si, J., et al. 2020. Review of artificial intelligence techniques in imaging data acquisition, segmentation and diagnosis for COVID-19. In *IEEE Reviews in Biomedical Engineering (Early Access)*.

Smith, M. J. 2018. Getting value from artificial intelligence in agriculture. *Animal Production Science* 60: 46–54.

Stetco, A., Dinmohammadi, F., Zhao, X., et al. 2019. Machine learning methods for wind turbine condition monitoring: A review. *Renewable Energy* 133: 620–635.

Touretzky, D. S. Building the Pascaline: Digital computing like it's 1642'. In *SIGCSE '15: Proceedings of the 46th ACM Technical Symposium on Computer Science Education*, Kansas City, Missouri, USA, February 2015, pp. 688.

Voyant, C., Notton, G., Kalogirou, S., et al. 2017. Machine learning methods for solar radiation forecasting: A review. *Renewable Energy* 105: 569–582.

Wang, W. and Siau, K. 2019. Artificial intelligence, machine learning, automation, robotics, future of work and future of humanity: A review and research agenda. *Journal of Database Management* 30(1): 61–79.

Warren, M and Pitts, W. 1943. A logical calculus of ideas immanent in nervous activity. *Bulletin of Mathematical Biophysics* 5: 115–133.

Widrow, B. and Lehr, M.A. 1990. 30 years of adaptive neural networks: Perceptron, madaline and backpropagation. *Proceedings of IEEE* 78(9): 1415–1442.

Wirkuttis, N. and Hadas Klein, H. 2017. Artificial intelligence in cybersecurity. *Cyber, Intelligence, and Security* 1: 103–119.

Yannakakis, G. N. and Togelius, J. 2018. *Artificial Intelligence and Games*. Springer Publishing Company, Switzerland.

Young, T., Hazarika, D., Poria, S., et al. 2017. Recent trends in deep learning based natural language processing. *IEEE Computational Intelligence Magazine*. 13: 55–75.

Yu, K., Beam, A.L. and Kohane, I.S. 2018. Artificial intelligence in healthcare. *Nature Biomedical Engineering* 2: 719–731.

CHAPTER 3

Essentials of Internet of Things

Design Principles and Architectures for Its Application in Various Domains

Hariprasath Manoharan
Panimalar Institute of Technology

Abirami Manoharan and Shankar T.
Government College of Engineering – Srirangam

Yuvaraja T.
Vrije Universiteit Brussel

Dinesh Kumar
OSI Soft Australia Pvt Ltd.

CONTENTS

DOI: 10.1201/9781003121466-3

3.1 IMPORTANCE OF IoT

The evolution of Internet of Things (IoT) has changed the society in different aspects over the past 20 years. Since it is possible to integrate all objects with an intelligent sensing device, a high speed internet connectivity is needed for monitoring all devices [1]. Therefore, IoT will be considered as one of the biggest prospect for industries and to the entire society. When objects are connected with IoT then information will be shared with users who request for it. The major principle is to recognize the nature of object and to sense the behavior of the object so that proper functioning is guaranteed. In addition, there is no need for an individual to visit a particular place for monitoring the devices. Instead IoT will be integrated for monitoring all objects at the remote location itself. This provides a great advantage to users because this results in the management of devices at a very low cost. Even the growth of fifth generation networks will provide much support to society in different applications as follows,

   i. Agriculture

  ii. Medical diagnosis

 iii. Transportation

 iv. Grid management and

  v. Home monitoring

For all the above applications, different types of sensors will be integrated with a unique Radio Frequency Identifier (RFID) and all devices will be connected using a Bluetooth module. Even in many cases, an advanced wireless technology using Wi-Fi can be used instead of a Bluetooth module [2].

The step of IoT for monitoring different objects is shown in Figure 3.1. It can be seen from Figure 3.1 that the objects are monitored using six different steps which starts with the collection of existing data that is needed for monitoring in different applications. Once the data is fed to the sensing equipment then all old parametric values will be compared. Then the data will be connected to the internet for transmitting to users at receiver side. After connecting to an authorized user the intelligent device will sense the parametric values that are fed as input and reports whether the objects are functioning properly or not. In the next stage once the data is received by the user it will then be analyzed and if any uncertainty occurs then the user can examine and correct it from the remote location itself. Then, it will be combined with the support of things for secondary backup process. Final values are executed with human values and experience provided by the humans [3].

From Figure 3.2 it can be seen that the current status of integrating IoT in India is under development for various fields. It can be seen that

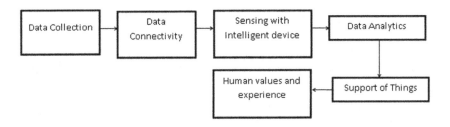

FIGURE 3.1   Steps of IoT monitoring process.

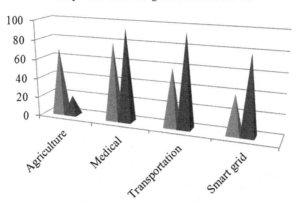

FIGURE 3.2   Current statuses on national and international level.

integrating IoT with different aspects is already developed in the United States. Even in India for both medical and transportation fields the application of IoT is much higher and is equal to the United States [4]. But the major backbone of India which is termed as agriculture is still lagging in the integration of smart intelligent devices. Therefore, from the current status it is clear that in all developing countries and in all applications IoT should be integrated so that all people can monitor from remote locations. Introducing IoT in all applications not only reduces people-to-people communication but it will enhance smart machine-to-machine communication. If machine-to-machine communication is introduced then it will provide opportunities for all entrepreneurs to start a new business model with more features in income. All devices which are used for monitoring the objects will not move toward granulose part, and therefore new type of services with uses and applications will be created. Moreover new types of services will be much feasible to all users. Further, India is mostly investing in digital market and it is evident that Make in India will only be possible by the integration of intelligent devices. By doing so, the economy of India is expected to grow higher and the knowledge of wireless devices to all people will also increase. But the question on security of wireless devices still remains for many people as the number of devices increase from millions to billions. As the number of devices increase, developing a safe high standard device at low cost is also possible by incorporating block chain technologies [5].

3.2 DESIGN OF AGRICULTURE FOR IoT

One of the important applications for IoT is agriculture where smart farming is required for monitoring and managing the crops. The information system that is used for managing the crops will be used for data acquisition process where IoT is integrated for optimizing business operations in agriculture. Figure 3.3 shows the design of IoT process for agriculture. From Figure 3.3 it can be seen that the design process of agriculture in IoT consists of a water sensor for checking and maintaining the level of water. This type of water sensor with conductivity level switch is very important because the level of water in field changes for every time period and to check the level of water at remote locations water sensors are preferred [6]. Also for checking the moisture content of soil a new type of soil sensor is introduced so that the type of crop to be planted in a particular field will

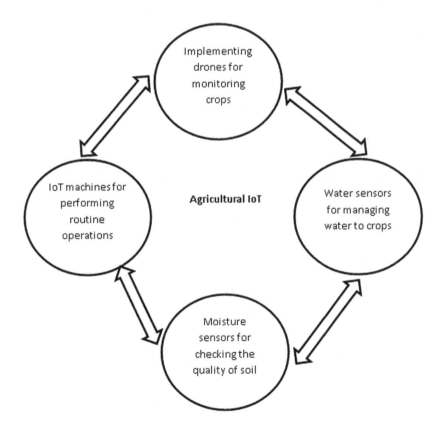

FIGURE 3.3   Design of agriculture for IoT.

be decided according to the quality of soil in the field. In addition, to prevent the crop from different diseases and insects it is better to use drones for monitoring the field [7]. If these types of advanced technologies are used then it is easy to monitor the agricultural fields and it will result in an increase in the productivity of crops [8].

3.3 ARCHITECTURE OF IoT FOR AGRICULTURE

The process of IoT in agriculture is a three layered architecture which consists of

i. Front end layer

ii. Gateway layer

iii. Back end layer

3.3.1 Front End Layer

The front end layer consists of n number of nodes that are connected with various sensors for both transmission and reception of data. Since different sensors are used it is very important to integrate corresponding nodes at front end layers as shown in Figure 3.4.

3.3.2 Gateway Layer

The central layer for communication process between data transfer and collection process is called as the gateway layer. In most cases, this gateway

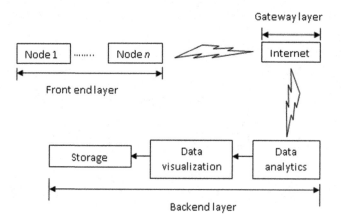

FIGURE 3.4   Architecture of IoT for agriculture.

layer will work with the zigbee module for data communication between transmitter and receiver.

3.3.3 Back End layer

The last layer of data communication process in the back end layer is where the process of data analytics and visualization occurs and separate API keys will be issued to all users who are currently retrieving the data at remote locations.

3.4 DESIGN OF MEDICAL APPLICATIONS USING IoT

Analysis of medical applications using IoT is usually divided into two different perceptions as follows:

   i. Engineering

   ii. Medical

IoT requires an engineering perspective where the problems can be solved by engineers with the help of designing implanted sensors or wearable devices that monitor the human body at all times. In most cases it is necessary to design low cost wearable devices such as wrist bands or rings which provide flexibility in monitoring different parameters of the body [9]. During such design process it is also important to have a low-weight communication device which is usually lesser than 30 grams. In addition, the applications on medical diagnosis using IoT need to be designed much more clearly than other applications because the system is going to monitor the health of all people in a continuous manner. Therefore, an advanced processor which uses a Wi-Fi module is needed in a single device with multiple sensors. This design process will be followed with the implementation of a wearable application-based system and an API security key should be provided to specialized doctors and pharmacists who are monitoring the corresponding patients. Implanted sensors can be considered only in occasional medical applications because the design process of implanted sensors consists of modules and signals that will affect the movement of the human body. Since the output of sensors is an electrical signal it is better to design wearables than implanting the sensors into the body [10]. Even though the size of implanted sensors is fine grains of sand more precautions and procedures should be followed in the

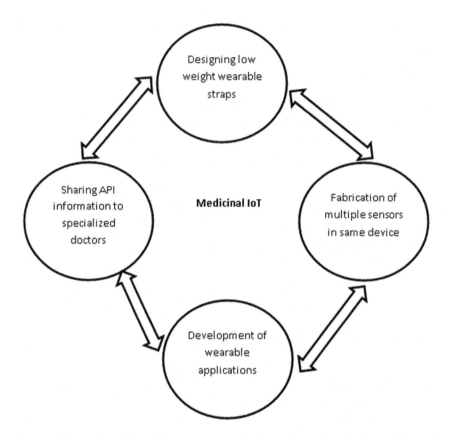

FIGURE 3.5   Design process of IoT in medical applications.

design process. Moreover, the power supply (Battery) that is produced for implanted sensors is not easy to replace inside the human body. Figure 3.5 shows the design process of IoT in medical diagnosis.

3.5 ARCHITECTURE OF IoT IN MEDICAL APPLICATIONS

The architecture of IoT in medical applications is different from other existing applications where a three tier approach is followed which consists of data generation, communication, and a relay node for collection of data [11].

3.5.1 Extremity Tier

This tier is used for generating the data that is aggregated from various IoT bases. Once the data is collected then heterogeneous data will be generated using various formats.

3.5.2 Intermediate Tier

In this tier communication between different sensors and nodes will take place through wireless modules. The information from this tier will be passed to a nearby base station and it will be reported to an ambulance in case of emergency.

3.5.3 Top Tier

Top tier is also termed as data collection process where all relay nodes are responsible for collecting the data from different sensors. The relay nodes which are connected with sensors aggregate all data from multiple sensors and send it to a diagnostic system through Internet.

From Figure 3.6 it can be seen that multiple sensors such as pulse oximetry, blood pressure and motion sensors are integrated in human body. All these three sensors will be connected to corresponding nodes with Zigbee protocol where a central monitoring unit will be associated with mobile application in either a mobile or a personal computer. Then, the information will be passed to a central access point which is an emergency center. Finally with an online monitoring system the observations will be plotted in analog form with waveforms so that specialized doctors can monitor the health of patients.

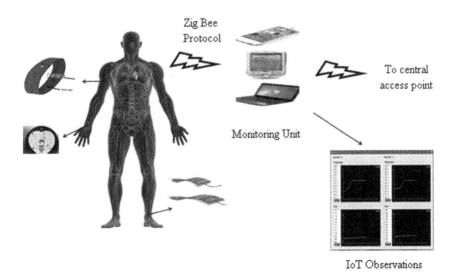

FIGURE 3.6 Architecture of medical application in IoT using multiple sensors.

3.6 DESIGN OF TRANSPORTATION PROCESS IN IoT

Recently, the application of IoT in transportation process is aggregated where automatic vehicles are created and manufactured. In automatic operation of vehicles more number of sensors will be implanted at both the front and back sides of vehicles [12]. In most of the vehicles if a speed wheel sensor is implanted then it is easy for monitoring the speed and if any fast movement of vehicles is detected then it will be reported to a traffic base station and therefore necessary action will be taken and an alert can also be given to the owner of the vehicle. The sensors will be designed to monitor the distance between two different vehicles at the front and back sides where a minimum of ten feet distance should be maintained. The main reason for introducing IoT in transportation application is to reduce the number of accidents that occur in a particular year [13].

In India during 2019 nearly 1.5 lakh people died due to road accidents and there was no monitoring system present in their vehicles for producing alert signals. Also due to drinking many people end up with road accidents which are very difficult to control by different forces. Therefore, it is necessary that the design process of IoT in transportation process is important for preventing accidents and if IoT is integrated in the transportation process the number of accidents will be reduced and safety will be guaranteed in transportation process. Even when the application of IoT in transportation process is introduced the high cost camera surveillance system can be replaced with low cost intelligent sensing devices such as sensors [14].

Figure 3.7 shows the design process of automobiles using IoT. From Figure 3.7 it can be seen that the design process of IoT in transportation is different when compared to other applications because in automobiles both front and back end sensors have to be designed and the designed sensors should be able to work under different temperatures at different vehicle speeds. The design process of automatic vehicle should be designed using the following four steps.

i. Enable maximum multipoint connectivity of sensors

ii. Afford flexibility to both front and back side of sensors

iii. Manage the lifecycle of vehicular sensors

iv. Check the quality of communication between front and back side sensors

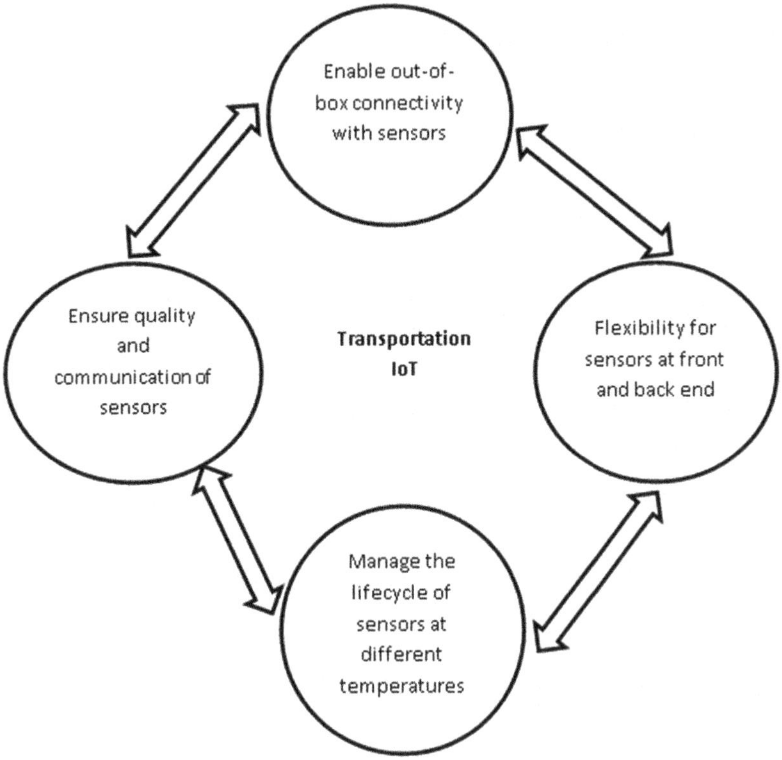

FIGURE 3.7   Design of automobile applications using IoT.

If all the four steps are properly designed then all parametric values can be easily monitored and the users can be informed about the operation of their own vehicles which in turn replaces the GPS system that is existing in vehicular applications.

3.7 LAYERED ARCHITECTURE OF INTELLIGENT TRANSPORTATION PROCESS

The architecture for transportation process should be designed in a distinctive manner where it should be able to track the passengers, goods, terminals, vehicles and junctions in and around the area. Therefore, for intelligent transportation process a four layered architecture is designed as follows.

   i. Application layer

  ii. Infrastructure layer

iii. Networking layer and

iv. Security layer

This four layered architecture usually consists of intelligent end points for tracking the required parameters. The end points will be in the form of sensors, GPS devices, mobile devices, surveillance cameras, logistic tracking and passenger flow detectors. The layered architecture should be able to support short range, vehicle-to-vehicle, fixed point and wide area communications. For fixed point communications a data center will be present for managing traffic, toll plazas, commercial vehicles, transit, fleet and freight services, whereas short range communications will be used for managing roadways and parking systems. Figure 3.8 shows the four layered architecture of an intelligent transportation system.

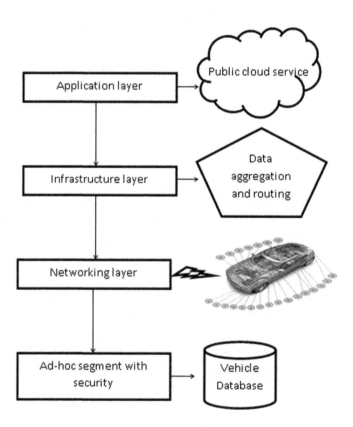

FIGURE 3.8 Architecture of intelligent transportation system.

3.7.1 Application Layer

The top layer in intelligent transportation process is the application layer where it consists of public cloud service for storing the data that is monitored by sensors. Here, the alert system will be integrated and if any unusual parametric values are detected the cloud based system will provide an alert to the server.

3.7.2 Infrastructure Layer

The collected data in the cloud is aggregated and routed to different vehicles in the wireless medium. Here, a new type of software system will be introduced at the traffic base station for receiving signals and to provide data on request.

3.7.3 Networking Layer

In the network layer the sensors which are implanted at the front, back and side of vehicles will be assimilated with a lightweight protocol for sending and receiving packets. The main advantage of networking layer is that variable length packets can be sent from one node to other nodes.

3.7.4 Security Layer

Since more number of vehicles are monitored a temporary work space should be created for data analysis. The creation of a temporary work space is to send ad hoc messages to users and to block any unauthorized attack by different users. Since ad hoc security is connected only authorized users can send or receive data in transportation process.

3.8 DESIGN OF SMART GRID APPLICATIONS USING IoT

IoT has advanced its applications in both transmission and distribution systems and it acts as an energy storage device. In a smart grid application the electric energy and information flows to energy systems and smart grid process. In addition a smart energy application includes charging of cars where a battery vehicle is operated with an energy storing device. The process of smart grid using IoT is an energy efficient process and it should be designed with a proper consumer maintenance procedure [15]. This type of application will offer great convenience to all consumers in both monitoring and control process. Many remote applications are designed for smart home applications for controlling the equipment in home. But when IoT is mainly applied in smart grid then bi-directional communication should be guaranteed and it is possible only with the help of advanced metering infrastructure. The old type of automatic metering systems are replaced by advanced infrastructure which performs the following tasks.

i. Management of demand

ii. Improving energy efficiency

iii. Alleviating the quality of power control and

iv. Saving energy

All the above functionalities will be performed using bi-directional communication using a central server system which connects with the help of data concentrators and automatic smart metering system.

The purpose of designing advanced smart metering is to install it in the locations of all customers which make the determinations such as calculation of monthly bills, power that is consumed by users etc. If in any case overload occurs then automatic turn "on" and "off" will be enabled which is considered as a diverse functionality that is integrated in the advanced metering scheme. Figure 3.9 shows the design process of smart grid which includes bi-directional operation with advanced monitoring systems [16]. The smart grid process is designed in such a way that demand response and self-healing should be more flexible in the distribution of power. If a flexible distribution operation is designed then consumption of energy will be reduced in electricity markets.

3.9 ARCHITECTURE OF SMART GRID PROCESS

The association of IoT in smart grid process is to check the capacity of all components inside the grid for transferring information either in wired or wireless medium. The architecture of IoT in smart grid consists of three layers as follows.

i. Information layer

ii. Power flow layer

iii. Power system layer

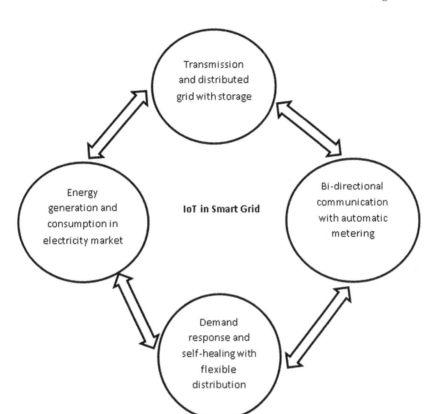

FIGURE 3.9   Design of bi-directional communication in smart grid.

Since the transmission system acts as a connector between both generation and distribution levels there is no prerequisite involved in introducing
the process. But for making the distribution grid smarter it is necessary
that all perilous points should be integrated with IoT frames. Figure 3.10
shows the architecture of a smart grid process with IoT. It can be seen
from Figure 3.10 that at the beginning stage during transmission and generation which is involved using a Local Area Network (LAN) the process
of IoT should be enabled, because if IoT is enabled during transmission the
problem of congestion management can be solved. Also another important aspect which relies on security can be resolved by integrating intelligent sensing devices with API keys. Therefore if intelligent sensing devices
are introduced then the information to a grid operator will be sent and all
losses and disturbances that are occurring during transmission and generation process can be addressed at the beginning stage itself [17].

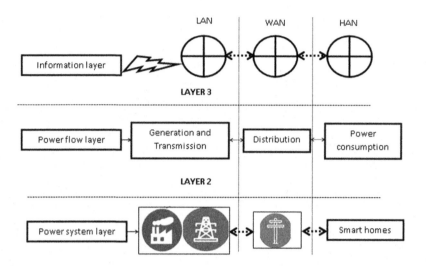

FIGURE 3.10   Architecture of smart grid process.

3.9.1 Information Layer

The information layer transmits the information in wireless medium and it is divided into three types depending on the requirement.

   i. Local Area Network (LAN)

  ii. Wide Area Network (WAN) and

 iii. Home Area Network (HAN)

Both LAN and WAN are used when a big grid network is employed, whereas for small buildings and homes HAN will be employed.

3.9.2 Power Flow Layer

The power flow layer is the important layer in smart grid process because this second layer is used for determining the type of power flow in the network and depending on the type of power flow the information layer allocated the corresponding networks. The power flow layer is divided into the following three types.

   i. Generation and Transmission flow

  ii. Distribution flow and

 iii. Consumption flow

From the aforementioned three types only consumption flow will have a small network process whereas the remaining two types will have a large network process.

3.9.3 Power System Layer

The lowest layer in any smart grid process is the power system layer where transmission and distribution grids are enabled. In addition the process of IoT will also be extended for smart automated home power systems which are connected with home network information flows.

3.10 DESIGN OF AQUACULTURE MONITORING USING IoT

The process of IoT is essential for monitoring aquaculture because it is one of the emerging areas in food sector and it is one of the important sources that supports sea plants and creatures to increase their lifetime through continuous monitoring [18]. Also, without IoT it is much difficult to monitor the aquaculture systems which are growing in seas and salty waters. The process of IoT reduces the involvement of workers who usually go inside the sea water for monitoring the lifetime of plants and animals. The process of IoT should be designed in a way where different sensors should be placed inside farm tanks and the incorporated sensors will be connected with a smart utility network [19]. The method of IoT which is involved in aquaculture is used for increasing the growth of fed farm animals and it will provide cost benefits to all farmers and aquaculture producers. All the fish farmers or business people who invested in aqua farms are much interested to know the status of farms at their remote locations. This is possible only with the help of IoT and proper and timely feed for aqua animals is needed to ensure their optimal growth during their life cycle. If all the aforementioned activities are monitored at the remote location then it will provide a benefit to the entire society and the integration of IoT will also be a cost-effective process [20]. Figure 3.11 shows the design of monitoring aquaculture using IoT where such a design using IoT will provide the following benefits.

    i. Minimization on consuming resources

    ii. Minimizing the food wastage

    iii. Minimization of impact on environment with aqua farms

    iv. Maximization of growth and life cycle of aqua animals

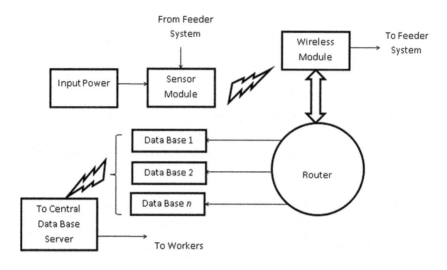

FIGURE 3.11   Design of aquaculture monitoring using IoT.

From Figure 3.11 it can be seen that the input power (+5 Volts) is provided to all sensor modules which is connected to the feeder system. The
monitored parametric values are passed in a wireless medium to a router
which aggregates all n databases' information to different systems. At the
receiver stage then database system will combine the information and forward it to a central database server where all workers can check the condition of aqua farms at their remote locations.

3.11 CONCLUSIONS

Since the necessity of appliances in day-to-day life is increasing in this
modern era it is essential that all modern contraptions should be integrated with Internet. Therefore, if a network is introduced as a data processing technique then it will be much easier to monitor all gears in remote
locations. This results in saving the time of individuals and highly accurate results will be distributed. If such technologies are enhanced then in
future it is much easier to develop a robotic technology where individuals'
work load will be much reduced. Also, this integration technique will have
reduced implementation cost which provides more advantage to all people
around the world.

REFERENCES

1. Javed, Farhana, Muhamamd Khalil Afzal, Muhammad Sharif, and Byung Seo Kim. 2018. "Internet of Things (IoT) Operating Systems Support, Networking Technologies, Applications, and Challenges: A Comparative Review." *IEEE Communications Surveys and Tutorials* 20 (3): 2062–2100. doi:10.1109/COMST.2018.2817685.
2. Patra, Manoj Kumar. 2017. "An Architecture Model for Smart City Using Cognitive Internet of Things(CIoT)." *Proceedings of the 2017 2nd IEEE International Conference on Electrical, Computer and Communication Technologies*, ICECCT 2017, 492–96. doi:10.1109/ICECCT.2017.8117893.
3. Zhang, Zhi Kai, Michael Cheng Yi Cho, Chia Wei Wang, Chia Wei Hsu, Chong Kuan Chen, and Shiuhpyng Shieh. 2014. "IoT Security: Ongoing Challenges and Research Opportunities." *Proceedings - IEEE 7th International Conference on Service-Oriented Computing and Applications*, SOCA 2014, 230–34. doi:10.1109/SOCA.2014.58.
4. Ayad, Soheyb, Labib Sadek Terrissa, and Noureddine Zerhouni. 2018. "An IoT Approach for a Smart Maintenance." *2018 International Conference on Advanced Systems and Electric Technologies*, IC_ASET 2018, 210–14. doi:10.1109/ASET.2018.8379861.
5. Chettri, Lalit, and Rabindranath Bera. 2020. "A Comprehensive Survey on Internet of Things (IoT) Toward 5G Wireless Systems." *IEEE Internet of Things Journal* 7 (1): 16–32. doi:10.1109/JIOT.2019.2948888.
6. Köksal, Omer, and Bedir Tekinerdogan. 2019. "Architecture Design Approach for IoT-Based Farm Management Information Systems." *Precision Agriculture* 20 (5): 926–58. doi:10.1007/s11119-018-09624-8.
7. Min, Byung Won. 2012. "Design and Implementation of an Integrated Management System for Smart Libraries." Communications in Computer and Information Science 310 CCIS: 186–94. doi:10.1007/978-3-642-32692-9_25.
8. Manoharan, Hariprasath, Adam Raja Basha, Yuvaraja Teekaraman, and Abirami Manoharan. 2020. "Smart Home Autonomous Water Management System for Agricultural Applications." *World Journal of Engineering* 17 (3): 445–55. doi:10.1108/WJE-07-2019-0194.
9. Boton, Conrad. 2016. "Challenges of Big Data in the Age of BIM: A High-Level Conceptual Pipeline." *Cooperative Design, Visualization, and Engineering* 9929 (November): 313–21. doi:10.1007/978-3-319-46771-9.
10. Budida, Durga Amarnath M., and Ram S. Mangrulkar. 2018. "Design and Implementation of Smart HealthCare System Using IoT." *Proceedings of 2017 International Conference on Innovations in Information, Embedded and Communication Systems*, ICIIECS 2017 2018-January: 1–7. doi:10.1109/ICIIECS.2017.8275903.
11. Sarkar, Sayan, and Renata Saha. 2017. "A Futuristic IOT Based Approach for Providing Healthcare Support through E-Diagnostic System in India." *Proceedings of the 2017 2nd IEEE International Conference on Electrical, Computer and Communication Technologies*, ICECCT 2017. doi:10.1109/ICECCT.2017.8117810.

12. Zhou, Hong, Bingwu Liu, and Donghan Wang. 2012. "Design and Research of Urban Intelligent Transportation System Based on the Internet of Things." Communications in Computer and Information Science 312 CCIS (70031010): 572–80. doi:10.1007/978-3-642-32427-7_82.

13. Hakim, Inaki M., and Aldeina Putriandita. 2018. "Designing Implementation Strategy for Internet of Things (IoT) on Logistic Transportation Sector in Indonesia." *ACM International Conference Proceeding Series*, 23–28. doi:10.1145/3288155.3288165.

14. Sun, Jianli. 2012. "Design and Implementation of IOT-Based Logistics Management System." *Proceedings -2012 IEEE Symposium on Electrical and Electronics Engineering*, EEESYM 2012, 603–6. doi:10.1109/EEESym.2012.6258730.

15. Saleem, Yasir, Noel Crespi, Mubashir Husain Rehmani, and Rebecca Copeland. 2019. "Internet of Things-Aided Smart Grid: Technologies, Architectures, Applications, Prototypes, and Future Research Directions." *IEEE Access* 7: 62962–3. doi:10.1109/ACCESS.2019.2913984.

16. Ghasempour, Alireza. 2019. "Internet of Things in Smart Grid: Architecture, Applications, Services, Key Technologies, and Challenges." *Inventions* 4 (1). doi:10.3390/inventions4010022.

17. Shahinzadeh, Hossein, Jalal Moradi, Gevork B. Gharehpetian, Hamed Nafisi, and Mehrdad Abedi. 2019. "IoT Architecture for Smart Grids." *International Conference on Protection and Automation of Power System*, IPAPS 2019, 22–30. doi:10.1109/IPAPS.2019.8641944.

18. Acar, Ugur, Frank Kane, Panagiotis Vlacheas, Vassilis Foteinos, Panagiotis Demestichas, Guven Yuceturk, Ioanna Drigkopoulou, and Aycan Vargun. 2019. "Designing an IoT Cloud Solution for Aquaculture." *Global IoT Summit, GIoTS 2019- Proceedings.* doi:10.1109/GIOTS.2019.8766428.

19. Balakrishnan, S, S Sheeba Rani, and K C Ramya. 2019. "Design and Development of IoT Based Smart Aquaculture System in a Cloud Environment." *International Journal of Oceans and Oceanography* 13 (1): 121–27. http://www.ripublication.com.

20. Manoharan, Hariprasath, Yuvaraja Teekaraman, Pravin R. Kshirsagar, Shanmugam Sundaramurthy, and Abirami Manoharan. 2020. "Examining the Effect of Aquaculture Using Sensor-Based Technology with Machine Learning Algorithm." *Aquaculture Research*, no. May: 1–11. doi:10.1111/are.14821.

CHAPTER 4

Theories of Blockchain

Its Evolution and Application for Security, Privacy and Trust Management

Navneet Arora

University of Liverpool

CONTENTS

DOI: 10.1201/9781003121466-4

4.1 INTRODUCTION

A very simple breakdown of the term Blockchain can be Block+Chain which means chain of blocks. It is a growing list of records referred to as blocks, where each block is linked to the previous block via a cryptographic hash value of the previous block.

Blockchain has been of interest to every researcher and technologist because of its unique design which not just prevents the modification of data but also acts as a distributed ledger while recording transactions in an efficient way so that it is verified and stays forever in record. It is significant that maintaining a ledger of data in distributed ways increases the applicability and probabilities of its adoption in various fields, as distributed computing allows access and management remotely (Iansiti & Lakhani 2017).

This chapter further discusses about the possibilities of its implementation, investments in blockchain and comparision of various earlier researches. The content provides basic understanding of from where blockchain evolved, what advancements it was introduced to, and what future it's going to make for the betterment of technology and mankind!

4.2 LITERATURE SURVEY

4.2.1 History of Blockchain: Its Evolution

4.2.1.1 Very Primary Stage

In the beginning of the 1980s, a cryptographer named David Chaum came up with a protocol which proposed a structure like Blockchain. His work was further carried forward by Haber and Stornetta in 1991 which illustrated the chain of blocks secured cryptographically. The motivation of this work was to implement a ledger where altering of timestamps would be impossible. Later in 1992, the efficiency of blockchain was improved when multiple certificates of documents were put into a block successfully (Sherman, Javani, Zhang, & Golaszewski 2019).

4.2.1.2 Primary Stage

Since 2008, there is one person whose name is also added along with Blockchain, whenever there is some discussion on Blockchain. The name is – Satoshi Nakamoto. It is also believed that it was not his individual work but the work of a group of people to conceptualize the very first blockchain. The cryptocurrency Bitcoin is highly famous with its name and this was the invention in 2008 by Satoshi Nakamoto as an implementation of blockchain. At this stage, there were two important enhancements in the design and structure of Blockchain as compared to the previous stage:

a. The use of a unique method to timestamp blocks without requiring the consent (signature) of trusted parties (nodes) in the network.

b. A new parameter to regulate the adding up of a new block in the chain (see Section 4.5).

This implementation was successfully done as the public ledger for all transactions on a network and was named Bitcoin (Nakamoto 2008).

4.2.1.3 Modern Stage

From 2008 to August 2014, the bitcoin blockchain size was increased to 20 GB which contains all the records (blocks) of a transaction in a chain that have occurred on the network. Then it gained the attention of common people and many across the world started investing and buying bitcoins. Since then, its size has been rapidly growing. It reached 30 GB in January 2015. In early 2020, It has reached more than 200 GB. Also, in the early times, Blockchain was referred as block chain but now it is referred collectively as blockchain (Statista 2021; Nian & Chuen 2015).

According to a research and customer service company in HR and IT-Gartner, by early 2018, 8% of high-level information officers showed their interest in short-term management and experimentation with blockchain (Gartner Reports, Artificial Lawyer 2018).

Many tech firms such as Accenture have been investing and researching in Blockchain. As per a report in 2016, Blockchain attained a rate of adoption as 13.5% which is often called as early adoption phase by people of market and management. Later in 2016, a lot of trade groups came together to start a global forum for the developments in Blockchain. This was mainly an initiative taken by the Chamber of Digital Commerce (Raconteur 2016).

4.2.2 Various Views on Blockchain's Applicability

There have been many diverse opinions of different researchers on blockchain's security, privacy and trust management services. However, the majority believes that blockchain has a high potential to achieve success (Table 4.1).

4.2.3 Basic Principles of Blockchain

Many authors have broken down this principles segment into a set of numerable principles but the most basic five principles which not just define the structural part but also illustrate some understanding about its working are as follows: (Janssen, Weerakkody, Ismagilova, Sivarajah, & Irani 2020).

TABLE 4.1 Related Work on Comparison of Various Views on Blockchain's Applicability

S. No.	Author & Year	Paper	Description	Advantages	Disadvantages
1.	Lansiti & Lakhani (2017)	The Truth about Blockchain	Blockchain has been viewed as the way of establishing contracts to regulate the relations between organizations, communities and individuals. They can serve as a medium to verify the events in organizations to provide security to assets while being a data structure to define legal and economic entities.	There is not much complexity provided by the transmission control protocol and Internet Protocol to adopt the blockchain because it is said that blockchain applications are being constructed above the infrastructure of data and communication.	Starting implementation from large scale is not advised because this might prove to be an inefficient approach if something fails. Also, only companies which are into financial services are ready to invest in the adoption of blockchain but manufacturing firms aren't much interested until now.
2.	Felin & Lakhani (2018)	What Problems Will You Solve with Blockchain?	Blockchain has already gained a lot of hype since the time it has evolved. If everything is planned carefully, then blockchain can gain an edge to eliminate the existing frameworks in various applications.	Replacement of existing frameworks by blockchain can result into streamlining the operations of core in various industries. It may also result in reducing the costs significantly and increased transparency in payments.	It is really required that managers should work upon the presentations to represent the ways in which their companies can be benefitted from adoption of blockchain. Each sector has its own different requirement from distributed ledger technology—blockchain.

(Continued)

TABLE 4.1 (CONTINUED) Related Work on Comparison of Various Views on Blockchain's Applicability

S. No.	Author & Year	Paper	Description	Advantages	Disadvantages
3.	Sherman, Javani, Zhang, & Golaszewski (2018)	On the Origins and Variations of Blockchain Technologies	David Chaum's vault system and all the early stages of blockchain have been explored to find out if blockchain will make certain future or not. More importance is given to compare the watchers, doers, executives and czars of various blockchain systems from Chaum's (1982) to Hyperledger fabric (2016), to find responsibilities of various roles in effective implementation of blockchain.	It is believed that blockchain addresses needs which are being inherited from many generations and because of its indelibility, it can be used from financial services to supply chain management sectors.	It is utmost to focus at the crucial characteristics of blockchain to examine the areas where it can be effectively implemented.
4.	Nakamoto (2008)	Bitcoin: A Peer-to-Peer Electronic Cash System	Bitcoin has a drawback of double spending which basically means using the same signature to verify a block more than once but a solution to this problem was proposed as Bitcoin—a blockchain based cryptocurrency.	A peer-to-peer network was researched and implemented using blockchain as the base technology. This network solved the problem of double spending by using the proof-of-work as a way to record history of transactions and verify new block to be added in chain.	Not many governments welcomed this because of the use of bitcoin in money laundering and other illegal funding involvement.

(Continued)

TABLE 4.1 (CONTINUED)   Related Work on Comparison of Various Views on Blockchain's Applicability

S. No.	Author & Year	Paper	Description	Advantages	Disadvantages
5.	Micheler & Heyde (2016)	Holding, Clearing and Settling Securities through Blockchain/ Distributed Ledger Technology: Creating an Efficient System by Empowering Investors	Both cost and risk of building and implementing this technological infrastructure of blockchain into banking depend upon the asset holders. There can be a group of regulators who should enable investors to form a common view regarding the risk associated with this technology.	This technology is easier to adopt with almost no barriers as this is a distributed ledger so many providers can be accommodated in this.	Transactions are made before sender and receiver have the cash and securities in place. So, blockchain technology does not appear to be in the efficient framework with the security concern. There are lot of discussions going on in this direction.
6.	Scott (2016)	How can Cryptocurrency and Blockchain Technology Play a Role in Building Social and Solidarity Finance?	Wide variety of factors have been discussed on subjects such as if blockchain can be used as a tool in financial inclusion, new cryptocurrencies such as Faircoin and blockchain 2.0 with its claims to be micro insurance and share issuances.	There have been attempts going on for creation of decentralized governance systems for individual banks to have voice within the decentralized technology systems.	It is believed that one blockchain does not fit all. The simple explanation is that all blockchain implementations in their respective domains will vary from each other due to different requirements.

(Continued)

TABLE 4.1 (CONTINUED) Related Work on Comparison of Various Views on Blockchain's Applicability

S. No.	Author & Year	Paper	Description	Advantages	Disadvantages
7.	Guegan (2017)	Public Blockchain versus Private Blockchain	Private blockchain works on a different philosophy than public which is based on digital trust. Nodes in private blockchain do not take part in public blockchain. The debate is still open on analyzing if the private blockchain are more fragile than public blockchain.	It is believed that the blockchain will fetch the credit for reduction in operational costs and significant increase in profit. This will also gradually result in the simplification of a lot of processes such as micro payments, micro credits for customers etc.	The debate is still open on analyzing the statement —"if the private blockchain are more fragile than public."
8.	Morabito (2017)	Business Innovation through Blockchain	There are three main areas on which emphasis is to be done, in order to develop a program based on blockchain: 1. Awareness on the existence of blockchain should be speedily spread. 2. Implementation of blockchain should provide the right services. 3. The connections in the ecosystem should be promoted.	The answer to the queries regarding blockchain becoming the fundamental in future can be given by the fact that bitcoin has achieved a 10 billion USD market worth and has also triggered other blockchain based projects because it is open-source to begin with.	A presence of suspicion will always be there regarding blockchain success because sometimes a lot of predictions go wrong and many such examples are also available.

(Continued)

TABLE 4.1 (CONTINUED) Related Work on Comparison of Various Views on Blockchain's Applicability

S. No.	Author & Year	Paper	Description	Advantages	Disadvantages
9.	Casey, Crane, Gensler, Johnson, & Narula (2018)	The Impact of Blockchain Technology on Finance: A Catalyst for Change	A lot of work is still required to be done, in order to develop more compelling use cases of blockchain but also some experimental study needs to be done on big data, machine learning and combination of digital innovations because without analyzation of such factors, results could be disruptive.	It has been concluded that blockchain possesses a real potential to act as a catalyst for bringing the changes in finances around the world.	Ultimate goal is to bring trust, which requires—an adequate framework of policy. For this banks, regulators, incumbents and disruptors are required to be brought together and discuss.
10.	Ganne (2018)	Can Blockchain Revolutionize International Trade?	Blockchain has been considered as trustworthy and highly resilient because it doesn't just serve as the base technology for cryptocurrencies like blockchain but it has a set of features such as tamper-proof, decentralized, distributed and digital ledger. These all features combined make blockchain highly reliable.	It is believed that blockchain can bring revolution to the international trade and take it to another level by breaking all the existent various barriers across cross-border trade transactions.	Blockchain is not believed to be suited for all purposes and situations. It can only serve in circumstances where multiple parties are playing a role in a transaction which requires high level of trust and transparency.

4.2.3.1 Distributed Database

Each party in a system with distributed database can have access to the entire database and also to its complete history of public ledger. In this type of system, no single party is given control to the data or to the information stored in the database but all the parties (nodes) in the network possess equal rights to access the database of blocks. Information is stored at each node and also forwarded to every new node added in the chain.

4.2.3.2 Peer-to-Peer Transmission

In blockchain, all the peers involved in the system can communicate with each other directly. There is no central node above the whole chain or network to provide the service of communication between any pair of nodes. All nodes involved in the system have the capability to verify or validate the records of their transaction.

4.2.3.3 Transparency with Pseudonymity

Details and records of each transaction are accessible to all nodes that are on the network of ledger. In blockchain, each node or user which belongs to the system has a unique 30-plus-character which is an alphanumeric address which helps to identify a node uniquely. If users want, they can remain anonymous else they can provide the proof of identity to other users in the system.

4.2.3.4 Irreversibility of Records

After each transaction, details regarding that are entered in the database; hence the accounts are updated accordingly. Once this is done, the records entered cannot be changed or altered in any way. The reason for this is that these records are linked to every other transaction which was done before them. That is why the term chain can be associated with blockchain. The records in the database have been made permanent, ordered chronologically and also made visible to others, who have access to the system. These functions are done by deploying numerous computational algorithms and approaches within the network.

4.2.3.5 Computational Logic

Transactions in blockchain are linked with computational logic and in this way transactions can be programmed. This shows the digital nature of the ledger. By this procedure, users or nodes in the blockchain can have algorithms and rules set up and deployed, which will help to trigger transactions automatically between the users or the nodes.

4.3 BLOCKCHAIN ARCHITECTURE

4.3.1 Working Process of Blockchain

To understand its working, one can think of a situation where some money/information is sent from sender(A) to receiver(B) in a more secure fashion and automated way, where the sender and receiver are supposed to be two nodes on the same network.

When a transaction is made, or we can say a record is created upon some transfer from A to B, it leads to the creation of a block.

Figure 4.1 shows that upon making a transaction, a block is created which doesn't hold much information except what is required, thus making it more secure.

After creation of the block, this block is then sent to all the nodes across the network for verification. Once all the nodes verify it, then it becomes a validated block which then qualifies to be added in the list i.e., blockchain. If any of the nodes in network fail to verify the block then the block is discarded and not added in the list. Thus, it becomes secure as it goes through verification by all nodes in a network.

A verified block acts as a digital notary. Each block which has been verified by all nodes is added to the chain and then the whole ledger (chain) is stored across the network in a distributed mode, so that any node can again access all transactional record/list of blocks. Thus, it is precise to say that a log of blocks/records is maintained by each node in the network.

Figure 4.2 illustrates that a block gets added into a chain upon verification from all the nodes. Forging or altering a single block would mean falsifying the entire chain in millions of instances of blocks which is not possible. Altering of a block is impossible as all the blocks are added up

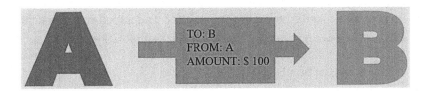

FIGURE 4.1   Creation of a block.

FIGURE 4.2 Verification at all nodes.

FIGURE 4.3 Falsifying a block.

in chain following a chronological order according to the computational algorithms. Also, each block in the chain is linked with the previous block though a hash value and changing the unique hash value of a block means falsifying the whole chain further.

Figure 4.3 illustrates that when an attempt was made to change the hash value of the second block, then it is not further linked with the consecutive block as the consecutive block holds the older value of the hash of the previous block in the chain. Thus, the second block's hash can't be forged, or we can say that a block in the chain can't be altered.

Thus, to summarize how a block is created and added in a blockchain on a network, we can simplify it as below (Nakamoto 2008):

1. A new transaction is created.

2. Each node collects the information about new transaction into a block.

3. Each node works upon verifying the transaction in the block.

4. Upon verification from all nodes, the block is accepted.

5. Validated block is added to the chain by working on creating the next block in chain, using the hash of the accepted block as the previous hash.

In the network, always the biggest chain is considered as the correct one and it keeps extending. If there arises a situation when two nodes broadcast different next block for addition in the chain, then other nodes may receive any one of them. In this scenario, work is done upon the first one but the other branch is also saved with a possibility that it might become longer. The nodes that were supposed to work on the second branch will switch to one of the branches depending upon which one is longer (Dobson 2018).

4.3.2 Types of Blockchain

4.3.2.1 Public Blockchain

In this type of blockchain, any user or node that has access to the network is able to read the whole record of transactions i.e., it possesses a capability to access all of the ledger, communicate to any node in network, verify and validate any new block/record which is to be added in chain. They are also able to create a transaction which becomes a block later—and gets added up in the chain.

In public blockchain, each and every node on the network is able to participate in consensus process i.e., the process of verifying and validating a block. This can also be understood as a process of deciding which block should be added to the chain and knowing the current state of that block. Like the very basic principle of blockchain—decentralized database, public blockchain is decentralized to the farthest extent.

4.3.2.2 Private Blockchain

In a completely private blockchain, the write authority, i.e., creating a transaction, is given to one organization. So, it will be able to alter the rules of blockchain i.e., changing up of computational logic, rules for verifying and validation. For example, transactions can be reversed etc.

Read authority, i.e., whole ledger access rights, can be made public or limited to a specific group of participants. This will lead to an increase in the privacy level in this type of blockchain.

4.3.2.3 Semi-Private Blockchain
Semi-private blockchain is owned by a company or organization. They have the right to grant access to the users/nodes who will qualify a criterion to be a part of the network. This type of Blockchain is used usually by business organizations. Government entities for storing the accounts, public records are the examples for this type of blockchain.

4.3.2.4 Consortium Blockchain
Consortium blockchain enables a set of nodes on network to take part in consensus i.e., rights to verify and validate the entry of a block in the chain are limited to a set of nodes. For instance, we can think of a group of 15 institutions where each one operates a node. For making a block valid, there must be ten approvals, so out of the 15 nodes only 10 will have the rights to verify and validate the new block which is to be entered in the chain.

Also, the read authority i.e., access rights can be kept public or can be limited to a specific group of participants/nodes (Figure 4.4).

4.3.3 Algorithms in Blockchain
The main functions of algorithms in block chain are as follows.

To verify signatures (hash values), confirm balances, perform validation of a block, determination of a way to instruct minors to how to validate the block, and to reach a consensus in the chain of blocks. In blockchain, there are mainly five consensus algorithms, out of which the first two are the most dominant and widely used in recent days (Narayanan, Bonneau, Felten, Miller & Goldfeder 2016; Nguyen & Kim 2018).

4.3.3.1 Proof of Work (PoW)
This is the most widely used, well-known, straight-forward, and useful algorithm. It has three main components which are transaction, the blocks, and the data miners who generate the PoW that mitigates through each block in the chain to validate each block. Also, it prevents denial of service from the distribution nodes. A block proposer is the fastest minor.

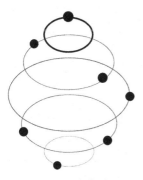

Public Blockchain

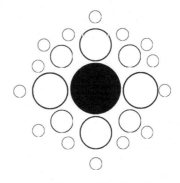

Private Blockchain

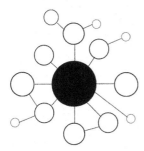

Semi-Private Blockchain

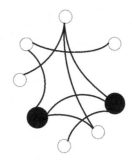

Consortium Blockchain

FIGURE 4.4   Types of blockchain.

4.3.3.2 Proof of Stake (PoS)

In this, a block proposer is the richest minor and it can be chosen in two ways:

a. via a selection of randomized block, it uses a formula which adds up the least hash value with stake of a block in chain.

b. via a coin-based selection, there are specialized nodes in PoS called as validators instead of minors. Validators are chosen on the priority of highest wealth (stake) in the corresponding asset for which they are validating.

4.3.3.3 Proof of Elapsed Time protocol (PoET)

It was made to function as a "fair" consensus model, which rotates around the concept of electing a leader in the distributed nodes of a network. PoET uses new, secure, CPU instructions which are rapidly growing among the customers of blockchain technology. Also, PoET uses CPU rather than GPU or ASICs. So, it creates a possibility for a bigger participation as it doesn't run on massive and expensive processors. Thus, it holds huge potential due to its capacity for allowing more nodes to serve as validators.

4.3.3.4 Proof of Importance (PoI)
It is an economic building protocol which was proposed by a blockchain organization called NEM. It is based on the concept of large account balances and trustworthiness; thus, it awards a score with respect to how reliable a validator is in the network.

4.3.3.5 Delegated Proof of Stake (DPoS)
It was designed to serve as the democratic and fair algorithm, and it is best among all the above illustrated ones. In this environment, employees (nodes) can change their boss (leader). In the DPoS environment, the stakeholders (elected delegates) can elect more than one "witness" (node) to generate next blocks in the chain. The process of consensus is known as approval voting. DPoS is not yet used widely because it is prone to malicious interventions of the stakeholders.

4.4 POTENTIAL USE OF BLOCKCHAIN IN VARIOUS APPLICATIONS

4.4.1 Healthcare
After 2016, there have been few significant implementations of blockchain technology as a base in the management of health records of patients. A data security firm in the Netherlands partnered with the Government of Estonia to create a framework solely based on blockchain to validate the identities of patients. It is called Guardtime. A blockchain based identity—Smartcard— was provided to the patients to keep a track of all healthcare events such as scheduling an appointment in database (Angraal, Krumholz, & Schulz 2017; Azaria, Ekblaw, Vieira, & Lippman 2016; Mettler 2016).

4.4.2 Education
Blockchain technology might change the way in which transcripts, degrees and certificates are generated. Also, information regarding the skills and experiences can also be stored and accessed on a network based on blockchain. Universities such as University of Nicosia and Massachusetts Institute of Technology have been exploring blockchain in

their operations. For example, certificates that are received by students from MOOC (massive open online course) platforms are being managed by using blockchain technology in University of Nicosia (Chen, Xu, Lu, & Chen 2018; Sharples & Domingue 2016; Skiba 2017).

4.4.3 Internet of Things (IoT)

The core features of Blockchain, such as distributed ledger, consensus algorithms, offers a suitability to deploy blockchain in IoT field. For example, IoT applications involving the management of sensitive data can use blockchain for hiding the real identities in the network. Also, Blockchain can be used in IoT for recording the data in a decentralized manner which will be owned by various stakeholders and thus it increases the trust (Ramachandran & Krishnamachari 2018; Dorri, Kanhere, Jurdak, & Gauravaram 2017; Danzi, Ellersgaard Kalør, Stefanovic, & Popovski 2017).

4.4.4 Gaming

There is a very popular implementation in games, CryptoKitties, which was released in 2017. This game has been successful in attaining a significant amount of traffic and congestion causing scalability issues on the Ethereum network because almost 30% of the transactions on the Ethereum network were caused due to this game (Barron 2018).

4.5 SECURITY, PRIVACY & TRUST MANAGEMENT OF BLOCKCHAIN

4.5.1 Security

As the blockchain is built and managed using a peer-to-peer network and a distributed timestamping server in which all nodes on a network play their role in verifying and validating a block, this design feature makes blockchain secure enough to be implemented in various applications such as international banking. All transactions are authenticated by a collaboration of massively powered nodes with their collective consensus or interests. A blockchain can even maintain a certain terms and rights in which each block will have to be accepted only after agreeing to the maintained terms and rights (Tapscott & Tapscott 2016).

Each block in a block chain can hold a set of validated transactions, which are already hashed and converted into a hash-tree. Each block is connected to the previous block via this cryptographic hash or the previous block. This preserves the integrity of the chain (Bhardwaj & Kaushik 2018).

4.5.2 Privacy

A private blockchain is also called a permissioned blockchain, and uses an access control layer to prevent everyone from accessing the network. Thus, a private blockchain ensures the maximum level of privacy that can be desired. All the validator nodes in a private blockchain are examined by the owner of network. There is also a disadvantage to this, which is that if one of the block creation resources is attacked then it might lead to halting of the whole block chain because in a private blockchain all block generation resources are centrally controlled by the network owner (Voorhees 2015).

4.5.3 Trust Management

The structure, design and implementation of Blockchain hold enough capability to address the major trust issues of global economists and investors in the field of Blockchain. It provides an immensely robust workflow where nodes in the network i.e., participants' doubts with relevance to the data security are marginal.

Due to decentralized principles, every node maintains its own copy of the whole blocks in the chain. Now the issue that arises is whether the data quality is equally good or not at all the nodes. This is also addressed by blockchain computational logics and data replication techniques which ensure that the data is nicely managed at all the nodes. All the users/nodes are equally trusted on the basis of the same data that they manage (Brito & Castillo 2013).

4.6 CHALLENGES AND ISSUES OF UTILIZING THE BLOCKCHAIN IN VARIOUS APPLICATIONS

In applications such as healthcare, there might be a few issues in the distribution of the highly classified health records within a public ledger. In a public blockchain deployment, each node has access to all the blocks, thus if blockchain is implemented in Healthcare with a view to increasing transparency then it might prove to be a disruption to the classified documents (Holotiuk, Pisani, & Moormann 2017).

Apart from this, it has been observed that when smaller blockchain-based applications have been implemented then they have also resulted in concerns related to the speed and scalability.

When it comes to implementing blockchain in IoT, then there is a list of issues such as there not being many standards and protocols available. Also, there have not been many documented best practices available to process the data which is collected by IoT devices or sensors (Kumar & Mallick 2018; Banafa 2017; Huh, Cho, & Kim 2017).

The general concern is that a simple blockchain implementation plan will not suit all the various domains of application because every industry has domain-specific requirements. Requirements of financial services differ from those of manufacturing.

4.7 SCOPE FOR TIGHTENING THE SECURITY, PRIVACY AND TRUST OFFERED BY BLOCKCHAIN SECURITY

4.7.1 Security

For maintaining a level of security in control and coordination of blockchain, one can set the block time during the implementation of blockchain. A block-time is the average time taken in a network to produce the next block which is to be added in the blockchain. So, this can be ensured that until a new block is generated, all of the information in the transaction of previously generated block gets verified.

Due to decentralized database in every type of blockchain, all the risks which are associated with the data held centrally in a particular location do not hold good. This decreases the vulnerability of attacks at a location to steal data by crackers.

4.7.2 Privacy

In a public blockchain, only those which are the nodes in the network possess:

  i. the rights to access the ledger

 ii. create a block

iii. participate in consensus.

Now, if there arises a need to put restrictions on a few nodes to perform any of the above tasks, then we have private, consortium and semi-private blockchain implementations. Thus, blockchain provides a high level of privacy depending upon which task is meant to be performed privately.

4.7.3 Trust

In an implementation of blockchain electronic cash system, double spending is a majorly observed issue. Double spending can be understood as passing same digital notary to all the nodes for verifying the transaction. In common words, it can be understood as duplication of a digital token in order to spend it more than once for verification. In blockchain, each node has to provide its own proof of work for verifying a block which is supposed to be unique for verifying a different block each time. Thus, blockchain removes this feature of infinite use of the same digital notary, solving the biggest problem of double spending in the eyes of investors.

Another trust issue which is effectively addressed is dealing with a situation when two or more blocks are produced simultaneously and are to be added in the chain. This creates a fork-like structure in the chain but this fork is actually temporary because algorithms of blockchain remove it quickly. The highest scoring branch of the fork is kept known and extended further.

4.8 FUTURE RESEARCH DIRECTIONS OF UTILIZING BLOCKCHAIN IN VARIOUS APPLICATIONS

Apart from security, privacy and trust benefits, blockchain has also been viewed as a technological base to minimize the cost and provides more efficiency. Therefore, it has been researched from the market point of view to invest less and make more profit.

In the recent times of technology, blockchain has been through much research and it is still being massively researched. It is believed that it might replace a lot of existing terms and technologies in various services such as banking, healthcare, data storage and processing techniques etc (Felin & Lakhani 2018).

There have been discussions going on in various forums, public domains and even at corporate levels to decide about its adoption and implementation at individual levels and at an organizational level. Scholars from the business and management world have been researching upon the various roles that blockchain can play to support its collabration with various existing technologies (Janssen, Weerakkody, Ismagilova, Sivarajah, & Irani 2020).

4.9 SUMMARY

With almost every trust issue addressed and every aspect of security and privacy covered, blockchain becomes a potential replacement for many existing technologies that lack high security levels and are vulnerable to data stealing attacks by crackers. As per a study conducted by Accenture, banking on blockchain could be more efficient, secured, and could serve $8 billion plus savings for just 8 banks. Thus, for global investors and economists of the corporate, IT, and service sector world, blockchain is a compromising solution for providing privacy, security, and trust management (Accenture 2017).

In the near future, we all will be able to witness its implementation in various sectors. It also might become a necessity to implement it as there will be emerging requirements to make systems more secure, private, and capable of handling many users and data. As we look for alternatives to provide all these quality services with attributes such as security and trust management, there is no other reliant technology than blockchain which is being researched highly in almost every technologically advanced country.

Furthermore, due to the variety of types in ledger providing various levels of security, it becomes possible to gain the trust of financial investors and create a possibility of its real implementation in near future. Hence, it is required to know from where blockchain evolved, what advancements it was introduced to, and what future it's going to pave for the betterment of technology and mankind!

REFERENCES

Accenture (2017) Banking on Blockchain. A Value Analysis for Investment Banks, https://www.accenture.com/_acnmedia/Accenture/Conversion-Assets/DotCom/Documents/Global/PDF/Consulting/Accenture-Banking-on-Blockchain.pdf.

Angraal, S., Krumholz, H., Schulz, W. (2017) Blockchain technology applications in healthcare. https://www.ahajournals.org/doi/10.1161/CIRCOUTCOMES.117.003800.

Azaria, A., Ekblaw, A., Vieira, T., Lippman, A. (2016) MedRec: Using blockchain for medical data access and permission management. *International Conference on Open and Big Data (OBD)*, August 22–24, 2016, Piscataway.

Banafa, A. (2017) IoT and blockchain convergence: Benefits and challenges. https://iot.ieee.org/newsletter/january-2017/iot-andblockchain-convergence-benefits-and-challenges.html.

Barron, L. (2018) CryptoKitties is going mobile. Can ethereum handle the traffic? https://fortune.com/2018/02/13/cryptokitties-ethereum-ios-launch-china-ether/.

Bhardwaj, S., Kaushik, M. (2018) Blockchain technology to drive the future. In: Satapathy, S.C., Bhateja, V., Das, S. (eds.) *Smart Computing and Informatics.* Springer: SIST.

Brito, J., Castillo, A. (2013) *Bitcoin: A Primer for Policymakers.* Fairfax County, Virginia: Geroge Mason University.

Casey, M., Crane, J., Gensler, G., Johnson, S., Narula, N. (2018) The Impact of Blockchain Technology on Finance: A catalyst for Change Geneva Reports on the World Economy 21. ISBN: 978-1-912179-15-2.

Chen, G., Xu, B., Lu, M, Chen, N. (2018) Exploring blockchain technology and its potential applications for education. *Smart Learning Environments.* 5(1). doi:10.1186/s40561-017-0050-x.

Danzi, P., Ellersgaard Kalør, A., Stefanovic, C., Popovski, P. (2017) Analysis of the communication traffic for blockchain synchronization of IoT devices, ArXiv e-prints, Nov. 2017.

Dobson, D. (2018). The 4 types of blockchain. https://iltanet.org/blogs/deborah-dobson/2018/02/13/the-4-types-of-blockchain-networks-explained

Dorri, A., Kanhere, S., Jurdak, R., Gauravaram, P. (2017) LSB: A lightweight scalable blockchain for IoT security and privacy, ArXiv e-prints, Dec. 2017

Felin, T., Lakhani, K. (2018). *What Problems Will You Solve with Blockchain?* MIT Sloan Management Review.

Ganne, E. (2018) *Can Blockchain Revolutionize International Trade?* World Trade Organization. ISBN 978-92-870-4761-8.

Gartner Reports | Artificial Lawyer (2018) Hype killer – only 1% of companies are using blockchain. https://www.artificiallawyer.com/2018/05/04/hype-killer-only-1-of-companies-are-using-blockchain-gartner-reports/

Guegan, D. (2017) Public blockchain versus private blockchain. https://halshs.archives-ouvertes.fr/halshs-01524440/file/17020.pdf

Holotiuk, F., Pisani, F., Moormann, J. (2017) The Impact of Blockchain Technology on Business Models in the Payments Industry, Proceedings of the 13th International Conference on Wirtschafts information, 912–926.

Huh, S., Cho, S., Kim, S. (2017) Managing IoT devices using blockchain platform. *2017 19th International Conference on Advanced Communication Technology (ICACT)*, Bongpyeong, pp. 464–467. doi: 10.23919/ICACT.2017. 7890132.

Iansiti, M., Lakhani, K. R. (2017) *The Truth about Blockchain.* Harvard Business Review.

Janssen, M., Weerakkody, V., Ismagilova, E., Sivarajah, U., Irani, Z. (2020) A framework for analyzing blockchain technology adoption: Integrating institutional, market and technical factors. *International Journal of Information Management.* 50: 302–309.

Kumar, N. M., Mallick, P. K. (2018) Blockchain technology for security issues and challenges in IoT. *Procedia Computer Science* 132: 1815–1823. ISSN 1877-0509, doi:10.1016/j.procs.2018.05.140.

Mettler, M. (2016) Blockchain technology in healthcare: the revolution starts here. *IEEE 18th International Conference on e-Health Networking*, September 14–16, 2016, Piscataway

Micheler, E., von der Heyde, L., (2016) Holding, clearing and settling securities through blockchain/distributed ledger technology: creating an efficient system by empowering investors. *Journal of International Banking & Financial Law*, 31(11): 11 JIBFL 631. ISSN 1742-6812.

Morabito, V. (2017) *Business Innovation Through Blockchain*. doi:10.1007/978-3-319-48478-5.

Nakamoto, S. (2008) *Bitcoin: A Peer-to-Peer Electronic Cash System*. The Academic Press. https://Bitcoin.org/Bitcoin.pdf

Narayanan, A., Bonneau, J., Felten, E., Miller, A., Goldfeder, S. (2016) *Bitcoin and Cryptocurrency Technologies: A Comprehensive Introduction*. Princeton, NJ: Princeton University Press.

Nguyen, G. -T., Kim, K. (2018) A survey about consensus algorithms used in blockchain. *Journal of Information Processing Systems*, 14(1): 101–128.

Nian, L. P., Chuen, D. L. K. (2015) "A light touch of regulation for virtual currencies". In: Chuen, David LEE Kuo (ed.), *Handbook of Digital Currency: Bitcoin, Innovation, Financial Instruments, and Big Data*. Waltham, MA: Academic Press. p. 319. ISBN 978-0-12-802351-8.

Raconteur (2016) The future of blockchain in 8 Charts. https://www.raconteur.net/the-future-of-blockchain-in-8-charts/.

Ramachandran, G., Krishnamachari, B. (2018) Blockchain for the IoT: Opportunities and challenges. https://arxiv.org/abs/1805.02818.

Scott, B. (2016) How can cryptocurrency and blockchain technology play a role in building social and solidarity finance?, UNRISD Working Paper, No. 2016-1, United Nations Research Institute for Social Development (UNRISD), Geneva.

Sharples, M., Domingue, J. (2016) The blockchain and kudos: A distributed system for educational record. In: *Reputation and Reward*. Adaptive and adaptable learning. Cham: Springer, pp. 490–496. doi:10.1007/978-3-319-45153-4_48.

Sherman, A. T., Javani, F., Zhang, H., Golaszewski, E. (2019) On the origins and variations of blockchain technologies. IEEE Security & Privacy, 17(1): 72–77.

Skiba, D. J. (2017) The potential of blockchain in education and health care. *Nursing Education Perspective* 38(4): 220–221. doi:10.1097/01.NEP.0000000000000190.

Statista (2021) Worldwide Bitcoin Blockchain Size. https://www.statista.com/statistics/647523/worldwide-bitcoin-blockchain-size/

Tapscott, D., Tapscott, A. (2016) Here's why blockchain will change the world. https://fortune.com/2016/05/08/why-blockchains-will-change-the-world/

Voorhees (2015) It's all about the blockchain, money and state. http://moneyand-state.com/its-all-about-the-blockchain/.

CHAPTER 5

Advances in Robotic Systems

Design, Modeling, Development and Control Principles

Priyanka Dhuliya

Graphic Era Hill University

Sunil Semwal

Model Institute of Engineering & Technology

Piyush Dhuliya and Diwaker Pant

Tula's Institute

CONTENTS

DOI: 10.1201/9781003121466-5

5.1 INTRODUCTION

Robotic systems have always fascinated the imaginations of scientists, engineers, youth, and children. As shown in Figures 5.1 and 5.2, be it the humanoid robot Asimo, the Mars rover, the industrial robotic arm, or the robots of Transformers, all have created interest in the minds to develop new robotic systems which can help in getting the task accomplished.

Robotic systems can be defined in a number of ways but the most obvious one is that it is a goal-oriented device which can perform tasks by sensing and acting on the acquired data through sensors [1,2].

The primary function of a robot is sensing because only then it can get data for further planning and other operations. These sensors can address vision, hearing, smell, and touch to accomplish the tasks [3,4]. All the data acquired through these sensors is worked upon by mathematical models or machine learning algorithms.

The algorithms actuate the control systems to drive the motors in a way in which one wants the motors to work.

FIGURE 5.1 Asimo Humanoid Robot by Honda. (Photo by Maximalfocus on Unsplash.)

FIGURE 5.2 Examples of Robots: (a) Mars rover. (Courtesy of mars.nasa.gov.). (b) Industrial robotic arm. (Photo by Possessed Photography on Unsplash.). (c) Transformer. (Photo by Somchai Kongkamsri from Pexels.)

5.2 DESIGNING A ROBOTIC SYSTEM

According to all the text books the design of robotic systems starts with problem identification. So, the problem identification is the primary task which needs to be addressed before the regular process. The problem identification lets us know as to what type of an output we want to receive and for that, what type of robot we can plan. Suppose we want to create a robot that only picks up and places the objects at a defined place or on a conveyor. For this we know that a simple robotic arm can help. Now in our mind the design is clear that we want to create a robotic arm. Next thing is how the robotic arm is to be made, what should be the dimensions of the objects that the robotic arm can pick and place, the degree of freedom required by the robotic arm for its movement, the amount of payload that the robotic arm can withstand, and the material which is to be used for making the robot taking into consideration the working scenario [5].

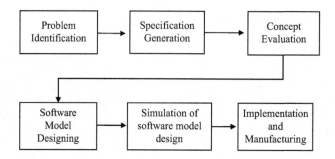

FIGURE 5.3   Block diagram of designing process.

5.2.1 Designing Process

As shown in Figure 5.3, the designing process starts with problem identification followed by generating specifications that need to be addressed, then concepts are evaluated to address the problem by taking into consideration the specifications, and finally a software model is designed. Earlier on prototypes were created but today software provides an entire simulation data of the conditions prevalent and the reliability of design. Based on the simulated data report and the design reliability report a decision is made for the manufacturing and implementation [6,7].

5.2.2 Designing Software

Robots are complicated as their electrical circuitry should complement their structure design. There are various software available in the market such as Solid Works, Auto Desk inventor, Fusion 360, Creo, Catia, solid edge, and AutoCAD. These software provide 2D and 3D modeling for the designing of mechanical parts so that one can get to know about the actual scenario before the real fabrication of robot. The designing gives us the precise dimensions for the cutting of metal sheets if a metallic robot is to be created or 3D printing is to be done. We can check the feasibility and the reliability of the design before the actual hardware through simulations. There are various simulation software available such as Webots, Gazebo, V-REP, Robotstudio which can make our task easier [8].

5.3 MODELING AND DEVELOPMENT

In this era, enormous interest in self-directed industrial vehicles application has been noticed. The system is designed in such a way that it can take all the aspects including the disturbance in their account. Their control systems should react quickly and adapt to prevalent environmental conditions and their changes. If we are considering designing a control system then methods such as algebraic control, optimal control etc. are utilized but for the robots their physical structure description and complete parameter description is required. A dedicated dynamic model is needed for the accurate description of mobile robot system. Properties such as torque, force, mass, inertia, frictional forces etc. should be taken into consideration for designing of a dynamic model. The structure and subsequent working of the robots can be understood by using such dynamic models which happen to be controlled mechatronic systems in totality. It is very important to create such a model for the development of highly complex systems. Here we are taking into consideration the physical modeling of the moving robot with two wheels and a castor wheel.

A model of robot elements is utilized to improve upon its way planning, navigational framework and confinement modules. We can also detect very early if there are any flaws in the robotic structure with the help of dynamic modeling. If there is any modification that needs to be implemented in the robotic system then it is easier through modeling as compared to the physical proto types which prove to be more costly as well.

Mobile robot dynamics and its knowledge is extremely important for both planning and development of feasible trajectories for a mobile robot, for this first a path planning module is considered which consists of a series of poses. For this a path is generated such that the mobile robot can traverse freely in the planned path without any hindrance to its movement, due to its dimensions dynamic properties are not of much importance at this time. The planned path happens to be merely wiped out geometrical spaces and the criterion for its planning is simply that any sort of stationary obstacle is not hit by the moving robot itself. Typical path planning algorithms for such applications are the A* and D* algorithms [9]. The moving obstacles in mobile robot workspace change things in the new scenario. There might be people, other mobile robots, or other moving machines. In this situation, the mobile robot must reach a particular pose during a fixed time-frame to avoid any collision with other moving objects. In this scenario, the velocity of mobile robot is to be constantly varied in a best possible way such that any collision between the moving object and moving robot can avoid any collision. In this way an accident free motion of moving robot and moving obstacles can be ascertained in the workspace [10].

During trajectory or path planning phase, dynamic properties of mobile robot are used to create a velocity profile group which will be utilized by the moving robot for maneuvering safely. The process involves limitations of velocity for mobile robots which should adhere to the trajectory which is planned [11]. The dynamic model of moving robot is very important when its velocities generate influencing forces which cannot be negated in the course of its movement. Moving robot soccer can be considered as a good example where velocities of small mobile robots are quite significant if we compare their size and their mass. The standard shape of mobile robot in soccer can be considered to be of a cube which has velocity more than 2m/s and size of approximately 7.5cm. The drive wheel may slip and turn over in the curved path due to a combination of small size and fast velocity. Slip of wheels in mobile robots can surely happen when there are enormous changes in velocity (big acceleration or deceleration values). So, appropriate control strategies and trajectories are to be included in the properties of mobile robots [12].

In [13] the authors devised a control strategy which involved a two-level mobile robot motion to deal with different workplace floor characteristics. The low-level part may be a classical controller for wheel velocity while the high-level part uses the measured velocities of wheel for adapting trajectory if there are major differences between the measured and desired values. Due to this the robot can be modeled as a hard-cased body that moves on one castor and two drive wheels which include the actuators for subsequently producing changes in velocity.

A dynamic model for moving robot is required for improving the estimated pose within the localization module. The kinematic model is mostly used for the prediction of pose for mobile robots using proper control input values. The control input values consist of rotational and translational velocities. Additional sensor measurements are used to correct anticipated pose. The use of non-linear Kalman filters can be considered as an example for this task [14]. The accuracy of the kinematic model, additional sensors and workspace model determine the quality of estimated pose. The kinematic model's accuracy features a substantial influence on such a structure. Control input for the kinematic model is ascertained when a low level microcontroller is connected with a navigational computer through a communication link. Both the velocity control of the driving wheels and their current velocity measurement are handled by the microcontrollers. The navigational computer receives velocity measurements from microcontrollers which can be used for path planning computations or estimation of pose. Due to delay in communication link, dynamic properties of moving robot and used drive mechanical characteristics (i.e., influence of friction, backlash) such as factors, input control values of the micro controller differ from navigational computer, as drive wheels velocity control is used by microcontroller. The results predicted by the kinematic model suggest the movement of mobile robot, while the microcontroller does not receive movement commands yet or any prediction for the same [15]. So, a continuing prediction error exists that cannot be considered by means of restrictions in path planning or calibration. It may be good to model such features and include them as a part of the localization module within the pose prediction step. Also, if a trajectory is made by altering the time-space for the moving obstacle avoidance part, path planning might be improved. To try it, an examination of physical properties of moving robot is done with determination of their influence on interesting variables within the case of navigation of moving robot, the vital variable such as its pose which is predicted and estimated using velocities of driving wheel. An appropriate model should use input variables which are velocity references to supply current as velocity value on its output for the mobile robot. Localization and other modules can then employ these velocity values. This is the primary step for model creation [16]. Validation of the obtained model is the second step. Thanks to the danger of injury, it is not right to authenticate the model on a true moving robot in its executing environment. Simulation testing is a more preferable way where mobile robot experiments-based velocity data are used.

So, there are two approaches to model the dynamic properties of moving robot features including the stimulus of mechanical drive characteristics. The primary approach is based on making a physical model of a moving robot body and its components used for velocity control like velocity controller, motor, and gearbox [17]. The second approach is based on experimental fitting of recorded moving robot velocity data regarding reference velocity data [18].

5.3.1 Physical Modeling and Development

A differential drive two wheeled with a castor wheel robot is considered. In this each wheel is independently driven. The motion in forward direction is produced when at the same rate both wheels are driven; for turning right, the right wheel is driven at a lower velocity than the left wheel and

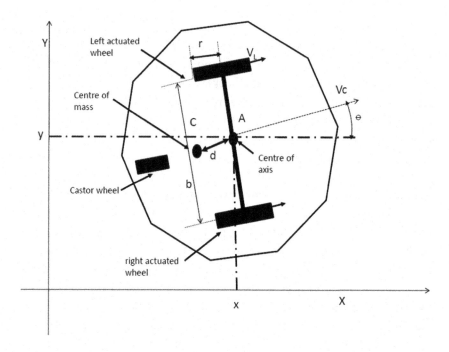

FIGURE 5.4 Geometrical dependencies of a differential drive robot.

similarly for turning left the left wheel is driven at a lower velocity than the right wheel. When one wheel is driven backward and the second wheel in its opposite direction at same velocity then the robot can turn at the very spot. The castor wheel which happens to be the third wheel provides stability to the robot. Driving wheels are armed with encoders which provide angular velocity readings through simple routine calls. A differential drive kinematic model for a mobile robot can be found in Section 5.3 [19] and its geometrical dependencies are given in Figure 5.4, where the drive wheel radius is given by r in millimeters, V_R and V_L are the velocities of the right and left driving wheel, respectively, in millimeters per seconds, x and y represent the position of the moving robot in Cartesian coordinates in millimeters, and b is the length of axle between the driving wheels in millimeters.

5.3.2 Model Implementation

The models can be implemented in MATLAB®/SIMULINK® or SCILAB®. Using certain requirements, the model can be implemented. The accuracy can be validated by the comparison of measurements obtained from a real moving robot and values of velocity for the model. For the comparison purpose, deviation of the estimated velocity from the measured one is computed including average and maximal error values.

5.4 CONTROL PRINCIPLES

Control is a very important element of robotic systems. There are different controlling techniques that can be utilized for robotic control. Today, much work is done on Fuzzy Logic Control, Artificial Neural Network (ANN), Deep Neural Networks (DNN). Although there are many more controlling techniques, we will limit ourselves to discussing the aforementioned control principles.

5.4.1 Fuzzy Logic Control

Fuzzy logic control involves the logic in which the analog inputs are treated as either 1 or 0. Robots contain many sensors whose inputs are to be processed so that a relevant output can be signaled to the actuators. Suppose the robot needs to overcome a simple obstacle, various ultrasonic sensors can be used which can measure the distance between a possible hindrance and the robot to avoid encountering the obstacle. For that various inputs will be given to the fuzzy logic controller which would be processing the given information to give an output to the actuators to move forward, turn right, turn left etc. to avoid the obstacle. The block diagram of fuzzy logic control for an obstacle avoidance robot [20,21] is shown in Figure 5.5.

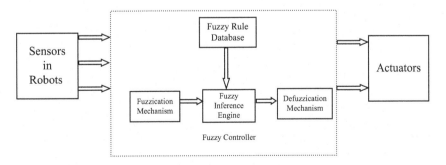

FIGURE 5.5 Block diagram for obstacle avoidance using fuzzy logic control process.

5.4.2 Artificial Neural Network

Artificial neural network (ANN) is a control system which emulates the nervous system. In this system we "teach" the system to provide suitable answers rather than programming the system. Therefore, it can provide relevant answers even when there is new information fetched to the system. In nervous system the fundamental unit is a neuron and the equivalent unit in ANN is the processing element as shown in Figure 5.6. The processing element combines using summations of input signals. The transfer function modifies the summed up input. The transfer function produces an output which may be working as an input to another transfer function. Weights are applied to the inputs when connecting to the other input paths [22]. This can be understood as providing synaptic strength to the biological neurons.

5.4.3 Deep Neural Network

Large artificial neural networks can be trained using deep learning. Deep learning can have many parameters ranging to millions which allow them to model non-linear dynamics. Deep learning covers the basic principles of regression. However, some advances have transformed the old regression into what is known as deep learning [23].

The use of machine learning for the control of robots requires them to leave some of the privileges of controlling them; at first it does not look good but looking at the overall benefits, the system can start learning on its own. This helps the system to adapt to the commands or directions which are communicated by the humans from time to time. Deep Neural Networks (DNNs) are very flexible and therefore they can be used with robots where

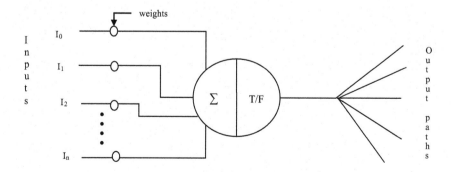

FIGURE 5.6 The processing element.

other machine learning models are not supported. Figure 5.7 shows some common structures where DNNs can be utilized for robots [24].

Figure 5.7a represents a regressing arbitrary function structure. For such types of robotic systems optimization techniques are implemented to reduce the prediction loss. Figure 5.7b represents an auto encoder structure. This model facilitates unsupervised learning. As can be seen it requires 2 DNNs: one is encoder and the other is decoder. In this only P needs to be given by the user while S is generated by the encoder itself. Auto encoders are very useful in robotics as sometimes it is difficult for the designer to figure out what values are required by the robot. In such cases the auto encoders perform the task themselves. This becomes extremely helpful when a combination of supervised and non-supervised learning is used [25].

Figure 5.7c represents a recurrent neural network structure which is used to model dynamic systems which include moving robots. In this P represents a control signal and may also contain some recent observations, Q represents an internal future state and S represents the future observations that are anticipated.

Figure 5.7d represents a structure that facilitates model free reinforcement learning. In this P is the representation of state and R is a control vector. S is the potential control vector.

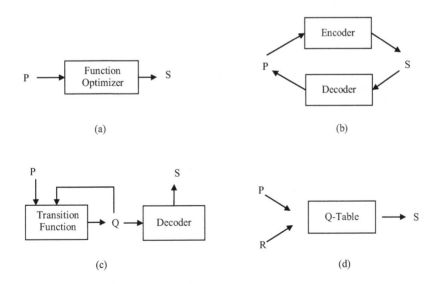

FIGURE 5.7   (a–d) Common structures for using DNN with robots.

This configuration is utilized when the task of the robot is known but the user does not know how to achieve it.

Various deep learning models are made up of stacked multiple layers of regression models. Among these models various types of layers have evolved for a variety of purposes. In those layers the convolutional layer has come up as a very important layer [26]. The convolution layer uses the same weights, unlike the fully connected layers, to operate across all the inputs space, due to which the total number of weights is significantly reduced in the neural networks. This is important with the images which have millions of pixels for processing. Therefore, convolution layers are important for the image processing tasks [27].

This has helped to ascertain robotic vision in certain applications such as gesture recognizing robots.

REFERENCES

1. Corke, Peter. *Robotics, Vision and Control: Fundamental Algorithms in MATLAB® Second, Completely Revised*. Vol. 118. Springer, Cham, 2017.
2. McKerrow, Phillip John, and Phillip McKerrow. *Introduction to Robotics*. Sydney: Addison-Wesley, 1991.
3. Rossi, Dino, Zoltán Nagy, and Arno Schlueter. "Soft robotics for architects: Integrating soft robotics education in an architectural context." *Soft Robotics* 1.2 (2014): 147–153.
4. Kanda, Takayuki, and Hiroshi Ishiguro. *Human-Robot Interaction in Social Robotics*. CRC Press, Boca Raton, 2017.
5. Craig, John J. *Introduction to Robotics: Mechanics and Control, 3/E*. Pearson Education, India, 2009.
6. Brooks, Alex, et al. "Towards component-based robotics." *2005 IEEE/RSJ International Conference on Intelligent Robots and Systems*. Albert. Canada, IEEE, 2005.
7. Bicho, Estela. *Dynamic Approach to Behavior-Based Robotics: Design, Specification, Analysis, Simulation and Implementation*. Shaker Verlag, Germany, 2000.
8. Bischoff, Rainer, et al. "Brics-best practice in robotics." *ISR 2010 (41st International Symposium on Robotics) and ROBOTIK 2010 (6th German Conference on Robotics)*. VDE, Berlin, Germany, 2010.
9. Carsten, Joseph, Dave Ferguson, and Anthony Stentz. "3d field d: Improved path planning and replanning in three dimensions." *2006 IEEE/RSJ International Conference on Intelligent Robots and Systems*. IEEE, Beijing, China, 2006.
10. Hoy, Michael, Alexey S. Matveev, and Andrey V. Savkin. "Algorithms for collision-free navigation of mobile robots in complex cluttered environments: A survey." *Robotica* 33.3 (2015): 463–497.

11. Haddad, Moussa, Wisama Khalil, and H. E. Lehtihet. "Trajectory planning of unicycle mobile robots with a trapezoidal-velocity constraint." *IEEE Transactions on Robotics* 26.5 (2010): 954–962.

12. Lu, Jianbo, et al. "System and method for dynamically determining vehicle loading and vertical loading distance for use in a vehicle dynamic control system." U.S. Patent No. 7,668,645. 23 Feb. 2010.

13. Dayoub, Feras, Grzegorz Cielniak, and Tom Duckett. "A sparse hybrid map for vision-guided mobile robots." (2011): 213–218.

14. Chen, S. Y. "Kalman filter for robot vision: A survey." *IEEE Transactions on Industrial Electronics* 59.11 (2011): 4409–4420.

15. Hamrita, Takoi K., Walter D. Potter, and Benjamin Bishop. "Robotics, microcontroller and embedded systems education initiatives: An interdisciplinary approach." *International Journal of Engineering Education* 21.4 (2005): 730.

16. Klančar, Gregor, and Igor Škrjanc. "Tracking-error model-based predictive control for mobile robots in real time." *Robotics and Autonomous Systems* 55.6 (2007): 460–469.

17. Rubio, Fernando, et al. "Comparison between Bayesian network classifiers and SVMs for semantic localization." *Expert Systems with Applications* 64 (2016): 434–443.

18. Kim, T. W., and J. Yuh. "Development of a real-time control architecture for a semi-autonomous underwater vehicle for intervention missions." *Control Engineering Practice* 12.12 (2004): 1521–1530.

19. Malu, Sandeep Kumar, and Jharna Majumdar. "Kinematics, localization and control of differential drive mobile robot." *Global Journal of Research in Engineering* Vol 14, No 1-H, ISSN 2249–4596 (2014).

20. Omrane, Hajer, Mohamed Slim Masmoudi, and Mohamed Masmoudi. "Fuzzy logic based control for autonomous mobile robot navigation." *Computational Intelligence and Neuroscience* 2016 (2016): 1–10.

21. Kharola, Ashwani, Dhuliya Piyush, and Sharma Priyanka. "Anti swing and Position control of single wheeled inverted pendulum Robot (SWIPR)," *International Journal of Applied Evolutionary Computation (IJAEC)* 9.4 (Oct-Dec–2018): 37–47.

22. Mohammed A. Hussein, Ahmed S. Ali, F.A. Elmisery, and R. Mostafa. "Motion control of robot by using Kinect sensor," *Research Journal of Applied Sciences, Engineering and Technology* 8 (2014): 1384–1388.

23. Sasaki, M., Suhaimi, M. S. A. B., Matsushita, K., Ito, S., and Rusydi, M. I. Robot control system based on electrooculography and electromyogram. *Journal of Computer and Communications* 3.11 (2015): 113–28.

24. Pierson, H. A., and Gashler, M. S. "Deep learning in robotics: A review of recent research," *Advanced Robotics* 31.16 (2017): 821–835.

25. El-Seoud, Samir Abou, Nadine Farag, and Gerard McKee. "A review on non-supervised approaches for cyberbullying detection." *International Journal of Engineering Pedagogy* 10.4 (2020).

26. Jha, Dipendra, et al. "Irnet: A general purpose deep residual regression framework for materials discovery." *Proceedings of the 25th ACM SIGKDD International Conference on Knowledge Discovery & Data Mining,* Anchorage, AK, USA. 2019.
27. Wang, Peng, et al. "Deep learning-based human motion recognition for predictive context-aware human-robot collaboration." *CIRP Annals* 67.1 (2018): 17–20.

CHAPTER 6

Data Science Practices

Running the Data Experiments Effectively for Statistical and Predictive Analytics on Data

Aatif Jamshed, Asmita Dixit,
Amit Sinha, and Kanika Gupta

ABES Engineering College

Manish Kumar

Krishna Engineering College

CONTENTS

DOI: 10.1201/9781003121466-6

6.1 INTRODUCTION

Transfer learning [1] is a significant contributor in every field of science, and one of the fields where image prediction needs development. This chapter explores how pre-trained models can be used in machine learning and the significance of transitional learning. Owing to the substantial dimensions of the SUV, road traffic control of the vehicle will be virtually impossible. However, some of the mechanisms (e.g. the clutch, brake pedal, and accelerator) are a little different than the hatchback. Thus, learning how to drive a hatchback would dramatically boost your ability to drive an SUV. You will benefit from the experience you already have from driving a hatchback as you learn to drive an SUV. The sedan model conforms with the adopted concept. To discover whether the image belongs to which animal, the author could use two different [2] methods. We will develop our own new learning paradigm to validate the methodology. Another approach to solve the issue is to use pre-trained deep learning algorithm that is currently being used to identify photographs of different animals like dogs and cats. By implementing the pre-trained model, this saves both time and money. Some benefits are derived from training a pre-trained model on real data. In all dog and cat pictures, some items such as flowers, leaves, the horizon and furniture appear. This pre-trained network is useful for recognizing any types of the items that never change or we can call them static or non-movable. A pre-trained network is thus a saved deep learning model, which has been trained on a very large data set, especially on image recognition problems. The model explains how to take different features, merge them together, and fine-tune them for running on a qualified network. This is what transfer learning does exactly. Transfer learning involves storing and utilizing information gained from one field of knowledge in order to learn in another similar area. To derive from pre-trained models, first use pre-trained models to classify the different images and then fine-tune to classify images of different objects. The paper offers a brief introduction to transfer learning and pre-trained neural networks as well as a reference to both. The proposed model uses VGG16 and ResNets to predict various images with varying success. Table 6.1 displays the VGG16 and ResNet50 comparison map [3]. The most reliable solution in the industry is to use our own deep neural network instead of an established convolutionary neural network (CNN). The new model's accuracy and training time are improved.

TABLE 6.1   Comparison between VGG16 & ResNet50

	VGG16	ResNet50
Accuracy	92%	96%
Epoch	25	25
No. of Layers	16	50
Optimizer	RMSPROP	SGD

In fact, however, you rarely write custom code for CNNs from scratch. You constantly retrain and change them to meet your requirements. The paper offers a brief introduction to transfer learning [2] and pre-trained neural networks as well as a reference to both.

The author will optimize our templates for optimal versatility. The author will clarify and analyze the neural network models that will be used later in Section 6.4. In planning to work with pre-trained models, one must first grasp the principle of transfer learning.

However, you almost never start with a completely integrated CNN in real-world ventures. To meet your expectations, you always adjust and train each puppy. This chapter would clarify the fundamental concepts of pre-trained models (used in the industry) [3] and how they can be transferred to a new application. The author will use images and, rather than attempting to create their own CNN from scratch, will instead use pre-trained models trained on the images to try to identify them. The author plans to make our models easier to manufacture. The author will use a set of unique models, which will be presented in this chapter. Before a researcher can begin training pre-trained models, one must grasp the principle of transfer learning [4].

6.2 LITERATURE REVIEW

F. Zhuang et al. [1] aimed to systematize current transfer learning results in a way that may help readers understand the processes and techniques of transfer training. To compare, this analysis analyzed over 40 descriptive approaches to transfers, in particular homogeneous approaches to transfers, from database and model perspectives. Pass learning is also briefly mentioned. At least 20 model-based classification models were used for studies to show the efficiency of previous transfer learning models.

G. Liang and L. Zheng [2] suggested a dilated convolution residual type for predicting photosensitivity in pediatric pneumonia. The research used an advanced diagnostic algorithm to diagnose correct pneumonia risks, and the low resolution, partial occlusion, and/or overlap of chest x-ray inflammatory areas could be solved efficiently.

J. Wang et al. [3] identified two new hybrid methods: (i) Manifold Dynamic Distribution Adaptation (MDDA) and (ii) Deep Transfer Learning Dynamic Distribution Adaptation Network (DTDN) (DDAN). A large amount of research has shown that MDDA and DDAN significantly improve the efficiency of transfer training and create a strong baseline against current deep and difficult digit recognition, feeling analyses and image classification methods. It has been shown that the marginal and conditional distributions relate differently to domain variations, and our DDA is able to provide a rational quantitative measure which results in improved efficiency and accuracy in prediction. They hope that this study would contribute to more analysis.

L. Torrey and J. Shavlik et. al. [5] addressed learning patterns, their goals and challenges. It summarizes the research on this subject, highlighting the current state-of-the-art and outstanding questions. The survey will be conducted in both inductive learning and reinforcement learning, and will explore negative transfer, as well as role analysis.

S. J. Pan and Q. Yang [6] compared and evaluated the effects of learning how to group, regress and cluster results. We will survey the relationship between transfer learning and other related machine learning phenomena such as domain adaptation, multitask learning and sample selection bias. In addition, we address potential future challenges for more efficient learning methods.

M. E. Taylor and P. Stone [7] developed a classification system for transfer learning and presented a framework that classifies transfer learning methods in terms of their capabilities and goals, and then use it to survey the existing literature, as well as to suggest future directions for transfer learning work.

6.3 PRE-TRAINED SETS AND TRANSFER LEARNING

Humans benefit from their experiences. If you want to become a decent driver, get an SUV. You have no experience of driving a larger vehicle. The SUV-sport utility vehicle or any other big car is longer than the hatchback, so it will be difficult to navigate in traffic. Yet, the basic structures of the hatchback remain the same in the eco system. Knowing how to drive a hatchback is a crucial skill to master while learning to drive an SUV. The basic driving skills you learned while driving a hatchback can be extended to any other vehicle type.

This is the transfer learning [8] principle. Transfer learning is the method of learning knowledge from one activity and immediately applying the knowledge to a similar activity. The hatchback-SUV model ideally fits the intended market concept.

The authors will have to recognize two possible methods in order to differentiate between various species. One is to construct a deep learning model from scratch and to supply the model with examples of photographs. Instead of building a deep learning model from scratch, a pre-trained deep learning model [9] able to identify cats has already been trained by utilizing cats' and dogs' images.

It saves time and money with the usage of a pre-trained model [10,11]. Using a pre-trained network often has benefits. All the pictures of small dogs and kittens are here. This chapter explores how we should classify items [12,13] such as trees, the sky, and furniture using this model. The usage of a pre-trained network [14] is a faster way to identify. Network users have to interpret and fine-tune full network configurations.

It is important to review the process of a CNN in order to fully understand feature extraction.

Figures 6.1 and 6.2 show a complete CNN architecture [14]:

Starting now, we will break the model into two pieces. The first segment includes all but network called ANN, and the second has just ANN.

A typical feature of deep learning is that it is considered a convolutional network, while a classifier is often built.

The bottleneck is the use of the convolutionary bottleneck [15], and the classification device is adjusted in feature extraction. The classifier would then aim to transfer different classifiers over the convolutional layer, and

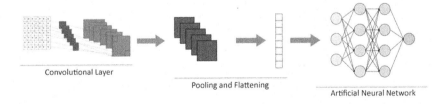

Convolutional Layer

Pooling and Flattening

Artificial Neural Network

FIGURE 6.1   Different layers of architecture.

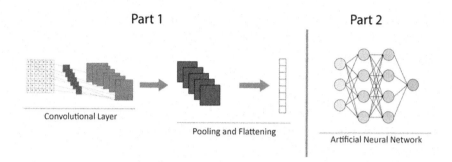

FIGURE 6.2   Convolutional base and classifier.

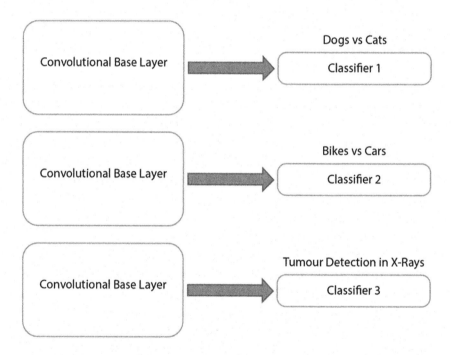

FIGURE 6.3   Classifiers with base layers.

the author will move different classifiers over the convolutionary layer. For example, a classifier may be a dog, a cat, a motorcycle against an automobile or even an X-ray diagnostic to classify a cancer, disease, etc.

Figure 6.3 [16] shows the base layers Cassfiers1, Cassfiers2, Cassfiers3

One interesting advantage of the convolutional layer is that since the convolutional layer can be trained for both inputs of interest, such as English and Chinese, we can obtain more general knowledge of the classification dilemma. To increase the misunderstanding rate of the classifier [17], we can only use the parameters of the samples to construct the model. You would also want to ensure that only the base layer is reused and not the casing.

The degree of generalization in a given layer has to be context based since each layer is different. In the case of cats, the network learns the basic characteristics of cats as an animal such as edges, while the subnet learns more specific features such as the edges of cats' paws, ears, and the outline of the nose. If you find an entirely different data set within your area of research, it would be better off if you did not spend all of your time on just adding on the different parts of the original data set—you're best off using a few large chunks of this original data set.

When pre-trained networks need to be updated, one of the most essential components is being able to try to further reduce the complexity of a network by freezing the layers above and below that they have already been trained on. In the "freeze mode," the network freezes the lower layers, so later the main vector values are trained from a previous session. Since the writers are using a pre-trained network, they will need to read the paper and pay attention to the initial layers, "Conv, DPNet, ResNet, VGG15 [18]." If we update the network with new data, it may be that we forget old data principles that the machine developed when processing. If the author of the paper uses a classifier (CNN) on top of the neural network, so each of the hidden layers will be randomly initialized, and the learning of the layers will be vanished completely.

A data collector must freeze certain layers to prevent data variability [19]. Making it unidentifiable or doing something to confuse the signature. When it comes to processes, we suggest that the method of fine-tuning is called preparation.

6.3.1 Adjustments in the Pre-trained Network (Tuning)

Fine-tuning means making small improvements to an inefficient or faulty process to correct the error. Fine-tuning deep learning involves the use of previous deep learning weights for programming a similar deep learning algorithm. Weights are used to bind neurons in each layer so that the network operates properly. The fine-tuning process allows for the deep learning algorithm to be performed more easily since it includes the basic details of the previous algorithm. Keras, ModelZoo, Torch, and MxNet are some of the most common fine-tuning machine learning libraries [20].

Fine-tuning means modifying our network so that the task at hand becomes more relevant. The author should backup a layer from the network that has no data to ensure there are no data loss. Detailed knowledge is essential and helpful. If the author is able to freeze those layers and unfreeze them, the author will reformulate some to make them fit the issue better. The author has a team that trained a classifier The author can make the layers like dogs and cats if they wished to. In this case, the network incorporated a new stage consisting of both dog and cat images. In this chapter, the author aims to perform an experiment and uses a pre-built network and trains it to recognize photographs of cats and dogs. There is a three-point method for adjustment or tuning.

1. ANN was integrated into a pre-trained computer .

2. The blueprint is frozen and constructed.

3. It is easier to train both the convolutional foundation and the added classifier jointly.

The following are the pre-trained networks that can be thought of as the base of convolutional layers. You use these networks and fit a classifier (ANN) [21]:

i. EfficientNetB0

ii. EfficientNetB7

iii. VGG16

iv. Inception V3

v. ResNet50

Figure 6.4 shows the top pre-trained models for image classification and performance.

The Microsoft Cognitive Toolkit (MTK), Inception V3, and Google MobileNets are examples of various deep learning tools/hardware that are built by different organizations to accomplish common goals. The next chapter will be on the VGG16 and ResNet50 versions, which are useful for learning to identify basic objects.

Model	Year	Number of Parameters	Top-1 Accuracy
VGG-16	2014	138 Million	74.5%
ResNet-50	2015	25 Million	77.15%
Inception V3	2015	24 Million	78.8%
EfficientNetB0	2019	5.3 Million	76.3%
EfficientNetB7	2019	66 Million	84.4%

FIGURE 6.4   Pre-trained model comparisons.

VGG-15 is a convolutional entwined neural network model with 15 layers that was recently proposed by Kasiviswanathan. Mr. Simonyan is a professor from Oxford University, and Mr. A. Zisserman is his classmate. In the L Sarkar et al. ILSVRC 2014 article, the AEC did some research of their Deep Belief Net (DBN) for a model that "came second" in 2014, ed. The AEC then went on to evaluate this DBN in other space-computed data sets, ed. This DBN will later go on to review the various visual features of a visual recognition framework. Instead of ResNet50, it may be applied to ResNet101 which is also regarded as the 1-layer ResNet CNN and refers to the network that was educated on the ImageNet data set that has 50 layers, earned first position in the ILSVRC 2015 competition.

6.3.2 Identifying an Image Using the VGG16 Network

The VGG model can be loaded and used in the Keras deep learning library [22]. Keras offers a library of neural nets which can be built and used as interfaces.

You can start off your modeling activities using the pre-trained weights given by the Oxford community and the neural network. Then, you can use the trained model to distinguish images, or otherwise, you can use it as a starting point of your own neural network.

The Keras backend supplies the 16-layer and 19-layer models of the model by class VGG16 and VGG19. Let's concentrate on the VGG 16 edition.

VGG 16 architecture, as shown in Figure 6.5, consists of a pooling layer, convolution layer, and a totally linked layer. The VGG network is an idea of deeper networks and narrower filters. VGGNet used a more complex neural network framework. Now, there are versions of 16–19 layers of VGG. These models held very tiny convolutional filters with $3\times3\times3$ taps. This implies that the filters were only looking at a small part of the

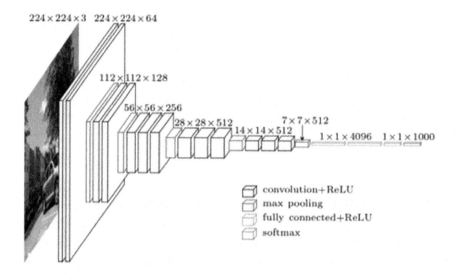

FIGURE 6.5 VGG16 network architecture. (From https://neurohive.io/en/popular-networks/vgg16/)

picture, thereby preventing us from overfitting. And they planned the network quite clearly with only 3×3 convolutional and periodic pooling at—layer.

A slice of pizza has an image of the writers. In the creation and recognition of photographs, the author uses the VGG16 network.

1. The repositories are imported.

2. Introducing the model.

3. Load the photograph. Load it.
    Figure 6.6 shows the predicted output by the pre-trained model.

4. Using img to array to change the image to an array:

5. In order to further processing, the image must be in a 4D shape for VGG-16. Extend the image dimension.

6. Plotting the image using the function pre-process input (Figure 6.7).

The author foretold in this exercise an illustration stating (with a likelihood of 97.68%) that the photo is a pizza. Of course, better precision means that in the ImageNet database [23] an object close to our image is present and our algorithm has found the image successfully.

FIGURE 6.6 Pizza slice prediction. (With permission from: https://www.shutterstock.com/search/pizza+slice+isolated+side)

```
array([[[[  -65.939    ,   -83.779   ,   -91.68   ],
         [  -72.939    ,   -89.779   ,   -97.68   ],
         [  -73.939    ,   -88.779   ,   -95.68   ],
         ...,
         [-101.939     ,  -114.779   ,  -121.68   ],
         [  -89.939    ,  -103.779   ,  -108.68   ],
         [  -66.939    ,   -81.779   ,   -88.68   ]],

        [[  -66.939    ,   -83.779   ,   -91.68   ],
         [  -71.939    ,   -89.779   ,   -95.68   ],
         [  -66.939    ,   -81.779   ,   -89.68   ],
         ...,
         [  -98.939    ,  -113.779   ,  -121.68   ],
         [  -72.939    ,   -88.779   ,  -100.68   ],
         [  -49.939003 ,   -67.779   ,   -75.68   ]],
```

FIGURE 6.7   Image pre-processing.

6.4 IMAGE CLASSIFICATION WITH RESNET

The Residual Learning Network is abbreviated as Resnet. The Resnet50 [23] has 50 layers. Resnet50 is a latent 50-layer network. Deep convolutional networks helped contribute to developments in image recognition. The trend is to extend the number of layers, as shown in Figure 6.8, so that we can solve complicated problems and improve precision. However, as we proceed with the neural networks, the accuracy begins to degrade. Planning will address this problem.

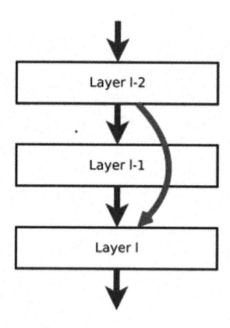

FIGURE 6.8 **ResNet layering.**

All these layers are used to process and condition the general deep neural network. Try to remember the forgotten, instead of mastering any of the functions. Rest can be removed as a feature obtained from the input layer. ResNet supports shortcut links (shortcuts from layer n to layer x). Training has shown to be easier than the training of simple, deep convolutional neural networks. The image of a car racer is used and model is trying to predict it through the network.

Following are the steps to complete this exercise:

Import the important libraries:

- Import library "numpy" as np.

- Import ResNet50.

- Initiate the build model and print the details.

- Load the image in the build model.

Note that the target size should be **224×224** since **ResNet50** only accepts **(224,224)**.

1. The code can transform the image into an array by using the img to array function.

2. The working picture should be in the shape of a four-dimensional sequence. Please let the dimension increase along the 0th dimension using the "expand dims" function.

3. Pre-process the input image by the script pre-process input.

4. To build the predictor variable for the image classifier, use it to predict by way of the predict process.

5. Check the features of the individual image.

6. Using the decode predictions feature to find the top five predictions.
   The first column contains an internal code number. Besides mark, the third thing is the chance of the object being labeled.

7. Then place the forecasts in a standard readable format. Display the most possible labels from the output of the decode predictions () function.
   In this respect, the model explicitly indicates that the image is that of a racer, as shown in Figure 6.9 (with a likelihood of 80.13%). That's the strength of pre-trained models and we can use and tweak these models by Keras.

6.4.1 Pre-trained Model Advantages

Previously, in the above article preceding this one, we looked at several different benefits of transferring learning in one way or another. Among other techniques, transfer learning allows us to build more robust models that can tackle a wide range of tasks.

- It helps to solve complex real-world problems with a number of limitations.

- If there are limited data to be used for preparation, there is a problem.

- Easy transfer of knowledge from one domain to another on the basis of learning tasks.

- It is a step toward the achievement of Artificial General Intelligence for the whole of humanity in the future.

FIGURE 6.9 Sample racer image for prediction. (From: https://www.mecomotorsports.co.in/wp-content/uploads/2020/02/racing-academy.png)

6.4.2 Pre-Trained Model Challenges

Transfer learning has huge potential and is often needed to improve existing learning algorithms. However, there are some significant questions in connection with learning transfer, which require more research and research. Apart from the difficulty of answering questions about when and how, negative transfers and transfers are challenging.

6.4.2.1 Reverse Transfer

We've seen several cases where transfer learning output will decrease. Negative transition happens where the shift of information from the source to the goal does not contribute to any change, but rather causes a decrease in overall results. There will be different explanations for transfer loss, such as not connecting the role properly, or whether the transfer process cannot exploit the interaction between the source and goal tasks very well. A thorough investigation is needed to prevent negative transition. Some authors demonstrate empirically, how brute-force conversion degrades output on tasks when the source and goal are too dissimilar. Bayesian methods by Bakker and their co-authors, and others studying clustering strategies, are being studied to prevent harmful transfers.

6.5 CONCLUSIONS

In this chapter, the author spoke about how transfer learning and its integration into neural networks could be the future of deep learning. The professional remembered the expertise he gained in the past, by using the pre-trained VGG16 and ResNet50 deep learning networks, to predict different photographs. At the outset, the author learned a pre-trained model by taking advantage of the techniques of feature extraction and fine-tuning to train the model faster and more accurately. Finally, the author found that they could use the same models, tweak them to fit with the data, and then use the data to develop the models. A very common strategy used by study groups is to take a pre-existing CNN (network of neurons) and then train it to identify new categories of inputs.

6.6 FUTURE WORK

Deep learning and neural networks are constrained by the machines they operate on. Deep learning utilizes machine learning to increase the performance of machine learning, maybe speeding up model creation. It will be awesome to see more pre-trained models that exploit this automation style of AI and business case studies which utilize it. Pre-trained models can effectively use natural language processing, most of the computer vision problems and complex problems of deep learning.

REFERENCES

1. Zhuang, Fuzhen, Zhiyuan Qi, Keyu Duan, Dongbo Xi, Yongchun Zhu, Hengshu Zhu, Hui Xiong, and Qing He. "A comprehensive survey on transfer learning." In *Proceedings of the IEEE Beijing 100190*, China, 2020.
2. Liang, Gaobo, and Lixin Zheng. "A transfer learning method with deep residual network for pediatric pneumonia diagnosis." *Computer Methods and Programs in Biomedicine* 187 (2020): 104964.
3. Wang, Jindong, Yiqiang Chen, Wenjie Feng, Han Yu, Meiyu Huang, and Qiang Yang. "Transfer learning with dynamic distribution adaptation." *ACM Transactions on Intelligent Systems and Technology (TIST)* 11, no. 1 (2020): 1–25.
4. Cai, Chenjing, Shiwei Wang, Youjun Xu, Weilin Zhang, Ke Tang, Qi Ouyang, Luhua Lai, and Jianfeng Pei. "Transfer learning for drug discovery." *Journal of Medicinal Chemistry* 63, no. 16 (2020): 8683–8694.
5. Torrey, Lisa, and Jude Shavlik. "Transfer learning." In *Handbook of Research on Machine Learning Applications and Trends: Algorithms, Methods, and Techniques*, edited by E. Soria, J. Martin, R. Magdalena, M. Martinez, and A. Serrano, pp. 242–264. IGI global, 2009.

6. Pan, Sinno Jialin, and Qiang Yang. "A survey on transfer learning." *IEEE Transactions on Knowledge and Data Engineering* 22, no. 10 (2009): 1345–1359.

7. Taylor, Matthew E., and Peter Stone. "Transfer learning for reinforcement learning domains: A survey." *Journal of Machine Learning Research* 10, no. 7 (2009): 1633–1685.

8. Apostolopoulos, Ioannis D., and Tzani A. Mpesiana. "Covid-19: Automatic detection from x-ray images utilizing transfer learning with convolutional neural networks." *Physical and Engineering Sciences in Medicine* 43, no. 2 (2020): 635–640.

9. Wu, Zhenghong, Hongkai Jiang, Ke Zhao, and Xingqiu Li. "An adaptive deep transfer learning method for bearing fault diagnosis." *Measurement* 151 (2020): 107227.

10. Naseer, Amina, Monail Rani, Saeeda Naz, Muhammad Imran Razzak, Muhammad Imran, and Guandong Xu. "Refining Parkinson's neurological disorder identification through deep transfer learning." *Neural Computing and Applications* 32, no. 3 (2020): 839–854.

11. Celik, Yusuf, Muhammed Talo, Ozal Yildirim, Murat Karabatak, and U. Rajendra Acharya. "Automated invasive ductal carcinoma detection based using deep transfer learning with whole-slide images." Pattern Recognition Letters (2020).

12. Chen, Yiqiang, Xin Qin, Jindong Wang, Chaohui Yu, and Wen Gao. "Fedhealth: A federated transfer learning framework for wearable health-care." *IEEE Intelligent Systems* 35, no. 4 (2020): 83–93.

13. Raghu, Shivarudhrappa, Natarajan Sriraam, Yasin Temel, Shyam Vasudeva Rao, and Pieter L. Kubben. "EEG based multi-class seizure type classification using convolutional neural network and transfer learning." *Neural Networks* 124 (2020): 202–212.

14. Jamshed, Aatif, Bhawna Mallick, and Pramod Kumar. "Deep learning-based sequential pattern mining for progressive database." *Soft Computing* 24(2020): 17233–17246.

15. Kumar, Arvind, Pawan Singh Mehra, Gagan Gupta, and Aatif Jamshed. "Modified block playfair cipher using random shift key generation." *International Journal of Computer Applications* 58, no. 5 (2012): 10–13.

16. Jamshed, Aatif, Surbhi Chandhok, and Romil Anand. "Analysis of sequential mining algorithms." *International Journal of Computer Applications* 165 (2017): 12–2017.

17. Jamshed, Aatif, and Garima Verma. "Mobile devices integration with grid by using efficient scheduling for local resource." *Journal of Advanced Computing and Communication Technologies* 1, no. 2 (2013): 1–6. ISSN: 2347-2804.

18. Aggarwal, Hari Shankar, Avishi Kansal, and Aatif Jamshed. "Noisy information and progressive data-mining giving rise to privacy preservation." In *2017 3rd International Conference on Advances in Computing, Communication & Automation (ICACCA) (Fall)*, Dehradun, India, pp. 1–5. IEEE, 2017.

19. The Deep Learning with Keras Workshop: An Interactive Approach to Understanding Deep Learning with Keras, 2nd Edition.
20. Raghu, M., Zhang, C., Kleinberg, J., & Bengio, S. (2019). Transfusion: Understanding transfer learning for medical imaging. In *Advances in Neural Information Processing Systems*, pp. 3347–3357.
21. Raffel, C., Shazeer, N., Roberts, A., Lee, K., Narang, S., Matena, M., … Liu, P. J. (2019). Exploring the limits of transfer learning with a unified text-to-text transformer. arXiv preprint arXiv:1910.10683.
22. Ruder, S., Peters, M. E., Swayamdipta, S., & Wolf, T. Transfer learning in natural language processing. In *Proceedings of the 2019 Conference of the North American Chapter of the Association for Computational Linguistics: Tutorials*, Minneapolis, Minnesota, pp. 15–18, 2019, June.
23. Tan, C., Sun, F., Kong, T., Zhang, W., Yang, C., & Liu, C.. A survey on deep transfer learning. In *International Conference on Artificial Neural Networks Rhodes*, Greece, pp. 270–279. Springer, Cham, 2018, October.

CHAPTER 7

The Actuality of Augmented Reality

Understanding the Realm of the Theoretical Framework for Embracing Its Potential Facilities

Ruchi Goel, Pavi Saraswat, Prashant Naresh, Anuradha Tomar, and Yogesh Kumar Sharma

AKTU

CONTENTS

DOI: 10.1201/9781003121466-7

7.1 INTRODUCTION

Augmented reality (AR) acts as a bridge between the actual world and virtual reality (VR) that can be done by using multiple sensors such as visual sensors, audio sensors, various object recognition sensors, and olfactory sensors [1]. AR also deals with the reaction of the system dynamically with the real-world scenario. It also gives 3D registration with respect to real-world objects, and provides information about how and what a system gets perceived from real-world activities in the digital world [2].

In AR technology, various sensors get embedded in the system to perceive the real environment like various websites which provide information to a person about how a product looks on him so that their purchase can be increased online.

Various hardware components required to implement AR in a system are processors, display, sensors, and cameras that are normally found in smartphones and laptops very easily [3].

7.2 BACKGROUND DETAILS OF AUGMENTED REALITY

In [4], it is explained that since AR is used in various industrial applications to make the whole system automotive in nature, AR maintenance is required in various aspects such as environmental, technical, and economical. However, developing a worker support system based on AR in industries is a highly challenging task.

In [5], a layout of virtual AR is proposed in which a virtual agent observes both virtual scenarios and real scenarios called as inverse VR. In this paper, the analysis and representation of real world in context of virtual agent is proposed.

In [6], an improved education learning application based on AR is proposed to make the education system more interactive with the system which, in turn, enhances the student's performance. In this application, all the teaching materials get represented in 3D images and videos.

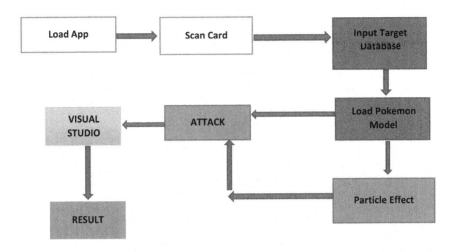

FIGURE 7.1   Flow to create advanced version of *Pokémon Go.*

In [7], a 3D game based on AR is proposed. Figure 7.1 shows the creation of an advanced version of *Pokémon Go* which embed fighting features that unite real world with the virtual one. In this game, they propose a system to load the app in the device and scan the card on it to begin the game. In this game, Vuforia database is used to store data and visual studio is used as front end to implement the game. In this game, the characters of the game popp up 3D characters which make the game more entertaining in nature.

In [8], a tourist information system formed on AR is proposed, in which beacon used to fetch tour information based on their point of interest. The information that is triggered by beacon-enabled smart devices by virtue of wave signal gets downloaded and matched with real images embedded in smartphones to visualize AR.

In [9], a recommendation system for tourists based on AR is proposed. This recommendation engine uses data of tourist attractions to visit and then provides the E-guide on their smart devices so that tourists will not face any problem while visiting a place.

7.3 BRIEF HISTORY OF AUGMENTED REALITY

In 1968, the very first Head Mounted Display was invented by Ivan Sutherland which gave way for various new developments in this field. In 1990, Tom Caudell who was the research scientist in Boeing coined the term **Augumented Reality** [10].

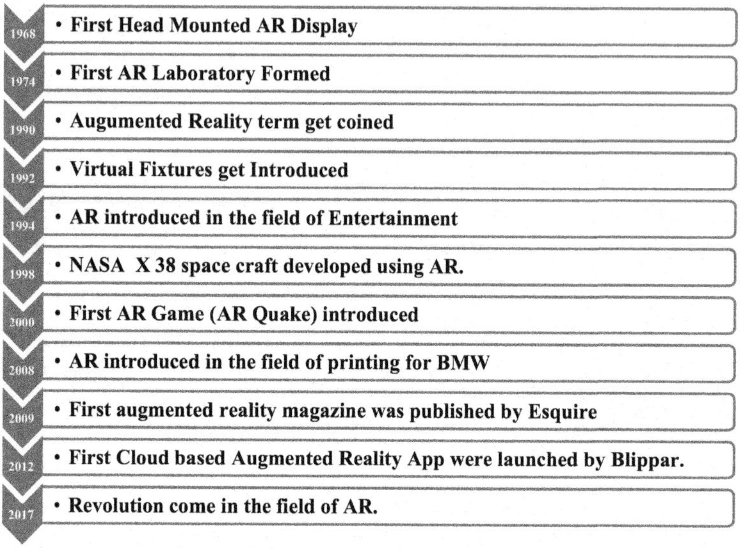

1968	• First Head Mounted AR Display
1974	• First AR Laboratory Formed
1990	• Augumented Reality term get coined
1992	• Virtual Fixtures get Introduced
1994	• AR introduced in the field of Entertainment
1998	• NASA X 38 space craft developed using AR.
2000	• First AR Game (AR Quake) introduced
2008	• AR introduced in the field of printing for BMW
2009	• First augmented reality magazine was published by Esquire
2012	• First Cloud based Augmented Reality App were launched by Blippar.
2017	• Revolution come in the field of AR.

FIGURE 7.2   History of AR.

In 1992, first Virtual Fixture was developed by L. Rosenberg which helped enhance the human performance in U.S. Army directly or via accessing various things remotely (Figure 7.2).

In 1998, NASA's X 38 spacecraft was developed which was based on AR. This spacecraft was designed to experiment the re-entry process during space missions [11]. The purpose of designing this craft was that if cataclysm occurs, then this craft was able to return them safely to earth. So, it was designed for automated navigation purpose in space missions.

In 2000, the very first AR shooting game known as AR-Quake was developed in which a player was able to see a total virtual environment via virtual goggles like a head mounted display by which the user can see both real and virtual world simultaneously. An AR-Quake game gamer can get the direction by GPS and orientation sensors that get synced with the real world. Monsters and whole architectural structure in the game appeared as a real world, and a player can play the game as he/she is in the game itself, i.e., in the virtual environment inside the game [12].

After this, a revolution occurred in the field of computer games where AR was used. In 2005, AR Tennis was launched in Nokia phone which was based on AR. In that, mobile phones' cameras were used with display and processing power in combination with computer graphics to create the real-world environment.

A boom occurred in the field of media in 2008 in context with AR. BMW was the first who used AR to print their enhanced AR-printed ads for their brand. With AR-printed ads, customers get more involved with the product and display so much interest than that in the traditional ads.

In 2009, the first AR magazine was published by Esquire in which readers used to scan the cover which made Robert Downey Jr come virtually on the magazine page.

In 2012, the first cloud-based AR app was launched by Blippar. They also developed the first AR game in 2014 that used Google Glass headset, which was unveiled at the Mobile World Congress.

In 2016, a very popular AR game was launched which was based on location by Niantic and Nintendo, named *Pokémon Go*. In this game, AR was put on standard map used in real life [10].

In 2017, a huge revolution was observed in the field of AR. The number of people using AR increased to 37 million. It is now expected that this number would increase to 67 million by end of 2021.

7.4 APPLICATIONS OF AUGMENTED REALITY

AR is an emerging field that allows us to interact with real-world objects. AR concept is utilized in various fields such as medical, archaeological, and commercial, as well as in designing [13].

7.4.1 Augmented Reality in Medical Field

1. **For finding veins of patients**: AccuVein is the start-up company which uses AR technology and helps nurses and doctors to find veins of patients. It involves a scanner that is projected over the body of the patient [14].

2. **Augmented reality-assisted surgery (ARAS)**: This technology helps surgeons to get an operational view by generating an image. The technique has been used in cardio and urological surgeries. Hence, using AR, the quality of surgery enhances while complexity reduces.

3. **In analyzing reports of patients**: Doctors can use AR to view real-time X-rays and CT scans in 3D to get more exposure and idea about

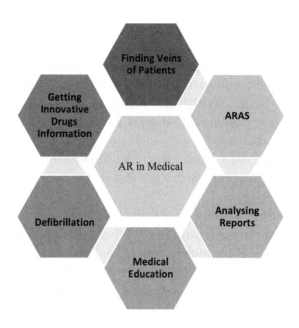

FIGURE 7.3   Various fields where AR is used in medical.

patient's disease. This involves constructing the images from different angles resulting in a more precise view.

4. **In medical education**: Students who are pursuing medical education
can now get a more advanced version of education. Many universities of the world are trying to give a virtual idea of internal anatomy
of human body. This will result in improved quality of education
(Figure 7.3).

7.4.2 Augmented Reality in Commerce

Many e-commerce sites and apps use AR for designing purposes and to
give their customers a more enhanced version. Most of the online websites
which are related to cosmetics and clothing gives the option of "try it now"
to their customers. Therefore, customers get a virtual idea about the product which helps in selecting the best for them. This feature also fascinates
the customers, e.g., Sephora which allows their customers to try on their
digital makeup. More examples of AR in retailing and in e-commerce are
as follows: L'Oreal, Sephora, Charlotte Tilbury, Shopify and Twinkl [15]
(Figure 7.4).

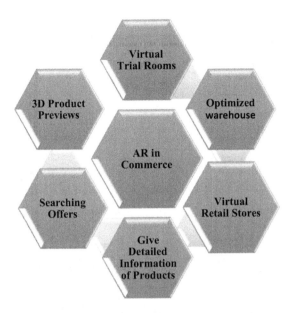

FIGURE 7.4　AR in commerce.

7.4.3　Augmented Reality in Archaeology

Archaeologists can use AR in their archaeological research. They can use AR for generating images of buildings and sites of ancient time which can help them in analyzing different parameters and configuration [16]. The technology enables them to interact and collaborate real world with the virtual world environment.

7.4.4　Augmented Reality Cloud

The **AR Cloud** represents real-time map in the 3D vision which represents the real 3D map onto the real world. It connects all the devices and apps physically located at various distances with each other and makes them to share and augment their information and experiences with each other [17]. Figure 7.5 depicts AR cloud architecture.

7.4.5　Augmented Reality in Fitness World

AR can be used in performance analytics of a person, how much they need to exercise in a day to shred extra pounds from their body. AR is used to measure the heart beat in real time so that one can know what is going on in their body which can help the person to stop from getting his body strained [18]. Figure 7.6 shows various fitness applications based on AR.

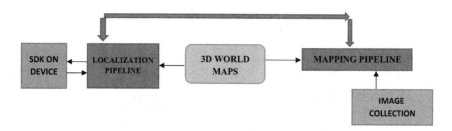

FIGURE 7.5   AR cloud architecture.

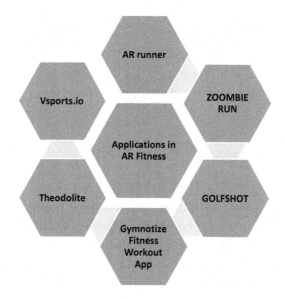

FIGURE 7.6   Various AR fitness applications.

7.4.6 Augmented Reality in Entertainment World

AR is emerging very fast in the entertainment world. It is getting very trendy to create motion pictures and various promotional videos [19]. Today, various music concerts are happening in VR and even various music lovers are paying extra charges to get entertained in VR.

7.4.7 Augmented Reality in Gaming Field

Various games have been developed using AR such as *Pokémon Go*. Since this game got very popular, developers were encouraged to develop more games based on AR [20]. In gaming zone using AR, the gamer can visualize the real world using virtual objects and characters, e.g., Temple Run, Treasure Hunt and many more.

7.5 CHALLENGES OF AUGMENTED REALITY

Overlaying of real-time computer animation or graphics onto the real world is known as AR. This term was coined by Tom Candell, an employee at Boeing, in 1990. From the time it was introduced it has come a long way until now. AR brings the virtual object to real world that we can work with. It allows merging of virtual information with the real environment to provide user with more immersive interactions with the surroundings. It can be stated that it is a more advanced version of the physical world.

But there are some challenges faced by AR as it is still in its early stage and remarkably new to the world and it will take some time to open out globally [21]. Figure 7.7 shows some of the challenges which are affecting mainstream acquisition of the technology.

7.5.1 Limited Content

One of the biggest challenges is *content availability*. Developers need to think about 3D content possible for this technology. So far, there are fewer number of applications based on AR technology. It is very hard, complicated, and super expensive to create AR content which can promote businesses [13]. Currently, the contents available with the AR technology consist of games or filters on social networking sites such as Instagram and Snapchat but there is no such application available in the market for which the consumer feels that urge to buy or use it in daily life.

This situation can be changed if more useful AR content is introduced by developers on which businesses could be built so that consumers could come in touch with the application and also it profits the business market.

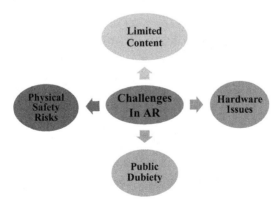

FIGURE 7.7   Challenges in AR.

7.5.2 Hardware Issues

AR uses large pieces of hardware that can be too expensive and the AR used cases are also of high cost which makes it difficult for common public to use. What causes delay are crucial hardware elements such as cameras, screens, motion sensors, processors and high-speed internet connectivity. On this front, we are still lacking by quite a distance [22]. And our smartphones do not have significant features to run a bulky AR application and so they are not user-friendly. Using AR on phones is inaccurate as the quality of picture is not proper. So, developers need to find a way to build cases which are cost efficient. Figure 7.8 shows the statistics of users who are using AR in various fields which make hardware challenge as a major issue.

7.5.3 Public Dubiety

AR technology is a very famous and hot topic among software developers and technical experts but public is *unaware about the uses and benefits* of the AR content. Consumers have only used AR technology in games or trying out furniture for homes or glasses on face [23]. But consumers do not have proper knowledge and have lack of awareness of the benefits it has in different fields. Users might feel uncomfortable to connect this technology with their accustomed life. The main cause of this is because it is too soon for AR technology and this concept is very new for the consumers and not many AR concepts are available to the public. A proper education is needed for the public to get aware of this technology and get comfortable with it.

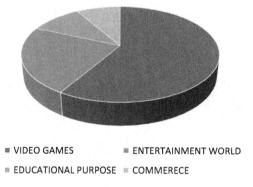

■ VIDEO GAMES ■ ENTERTAINMENT WORLD
■ EDUCATIONAL PURPOSE ■ COMMERECE

FIGURE 7.8 Users engaged in various fields of AR.

7.5.4 Physical Safety Risks

One of the major concerns is the lack of physical safety. It can cause physical harm when it is overlaid on the real world. The biggest example is *Pokémon Go* launched in 2016 and many were injured because of it [24]. Accidents and hazardous incidents can happen in risky surroundings such as construction sites and busy roads. The AR technology increases distraction which could be very harmful for people in daily life.

7.5.5 Security and Privacy Issues

Almost every technology faces the challenge of security and privacy problem. Without the permission and knowledge of the user, AR can be used in a wrong way. Hackers can access user's information and personal details and can *control the device of the user* remotely which may cause cyber threats [25]. If AR is used in GPS trackers and any hacker can take control to decide and change the directions, then it can lead to accidents and put our lives in danger.

Because of *lack of awareness* of the problems and technology, users do not understand what they are actually using and whether that application is safe or not which causes security and privacy issues. Figure 7.9 depicts basic requirements for security and privacy in AR devices.

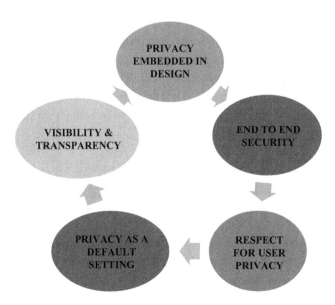

FIGURE 7.9　Basic requirements for privacy for AR devices.

7.5.6 Digital Fatigue

In the near or distant future as the AR spreads through worldwide and people become familiar with it, it may affect the mental health of people as they would be wearing AR glasses all day long with their vision partially obstructed which will make the line between real and virtual world unclear [26]. We can notice the impact of smartphones in our lives today, as we are glued to our screen most of the day which is affecting our health. So, the consumers and developers need to understand and spread awareness of the negative effect and harm in integrating AR to the world.

In conclusion: AR is so much new to the public and there are many challenges for AR which it needs to overcome. But we can take the example of Internet which was adopted in 1983 but the world took recognition of it in 1990 when world wide web was invented. So, in the same way, AR also needs time to get adapted in the globe and all the challenges would be vanquished.

7.6 FUTURE SCOPE OF AUGMENTED REALITY

As discussed in this chapter that AR is an emerging technology that enhances the users, experience of reality. The applications of AR in future are vast. Even in today's scenario too many business concepts are following medieval old techniques. One of the highly anticipated technologies that will improve the level of business is AR [27].

AR has applications in traveling, gaming, media, marketing, interior designing, architecture, education, entertainment, and others. Because of its vast application in different fields it is slowly increasing its implementation. The future for AR is vast as it has been proved in various statistics. It is expected that it will reach one billion users in today's scenario and by 2021, the number might increase if smart glasses running on AR software are increased to be approximately 5 million units.

The concepts of Internet of Things along with the AR technology will surely bring new fusion between the physical world and digital world. The future technical drift of using AR are as follows:

1. **3D Avatars**: This is going to evolve as an enhancement in the communication field. It forms a realistic 3D epitome of the user that imitates their actual expressions which in turn creates peculiar mesmerizing experience. This can be achieved by incorporation of already available applications such as Facebook with ARKit or ARCore.

2. **Next lineage in sales**: Embedding AR with traditional sales approach encourages the customer to simply visit the website of the product that they want to purchase and the sales representative will be there to join the user at his table.

3. **Gestural interfaces**: Manipulating the physical world without actually changing it using body language and action to control the technology.

4. **Visualization**: Visualization of the unlimited data present in today's technological time using the wearable technologies. It will find applications in various society matters such as law enforcement and emergency services.

5. **Future of marketing**: Marketing is like the backbone of any business; strong and effective marketing decides the future of a company and its products [28]. With AR in marketing, the advertisements can be made more and more interactive for the users. It helps the employees to create a positive impact of the products in users, lives. Users can view ads in a 3D view along with the description of the products in the presence of improved resolutions. This way the AR is expected to definitely increase its influence in marketing.

6. **Increasing number of users**: The number of users in AR is increasing according to latest statistics and studies [29]. In gaming field, the use of AR has tremendously increased for example *Pokémon Go*, Zombie run, Hogwarts Mystery and many more which introduced more and more users to AR, thus increasing the demand of more games based on AR tremendously.

7. **Social media platform**: Many social media platforms are building systems to accommodate AR [30]. According to the CEO of Facebook, AR is going to change the face of technology and also Facebook is working on a smart glass to incorporate AR and Facebook.

8. **AR, AI and ML**: Artificial Intelligence and Machine Learning are fast-growing sectors in the industry. So, bringing them together with AR and Mixed Reality systems is a natural extension of many of the things that are best suited to AI and ML, particularly computer vision. The ability to create human machine has immense potential by using AR.

9. **Remote assistance**: While using AR, engineers would be able to check the progress of the work on their sites using their AR-embedded devices in real time. They can also access data remotely and also send feedback to their subordinates while sitting remotely by establishing a connection with each other through the network.

10. **5G evolution**: AR is going to accelerate the 5G worldwide. AR integrated with 5G provides their users better experience with a low latency rate that could be assumed to be less than 25ms. 5G also fastens data transfer over cloud, and its processing and virtual image formation which in turn facilitate the user experiences toward AR.

11. **Automotive industry**: With the help of AR, the manufacturing and assembly unit of the automotive industry will be immensely benefited, as traditionally all the assembly and manufacturing processes are provided in a document form but with AR, this process performance gets improved using instinctual training and mandate potential that AR provides.

7.7 SUMMARY

This chapter summarizes about the actuality of AR and how it is affecting the real world. Initially information provided about the introduction to AR and how it is different from VR. Also, it briefs about the usage of the AR in disciplinary fields. Later, explains the background details of AR and where it has been already implemented and how it is changing the world for that matter. It also explains few proposals that can be initiated in this field by integrating AR with other fields. Afterwards depiction of the evolution of AR from initiation until now and how it grew in multiple fields is explained. It also explains the growth of the AR technology with time in multiple fields. Lately, talks about the applications of the AR in different areas such as Medical, E-Commerce, Archaeology, Fitness world, Entertainment world, Gaming field and in-depth knowledge of it and the major challenges faced by AR in multiple fields such as limited content issue, hardware issues, public dubiety, physical safety risks, security and privacy issues, and digital fatigue in detail. Finally, it explains the future scope of AR and the future possibilities in a detailed fashion.

REFERENCES

1. Azuma, Ronald T. "A survey of augmented reality." *Presence: Teleoperators & Virtual Environments* 6, no. 4 (1997): 355–385.
2. Billinghurst, Mark. "Augmented reality in education." *New Horizons for Learning* 12, no. 5 (2002): 1–5.
3. Bauer, Martin, Bernd Bruegge, Gudrun Klinker, Asa MacWilliams, Thomas Reicher, Stefan Riss, Christian Sandor, and Martin Wagner. "Design of a component-based augmented reality framework." In *Proceedings IEEE and ACM International Symposium on Augmented Reality*, pp. 45–54. IEEE, 2001.
4. Lorenz, Mario, Sebastian Knopp, and Philipp Klimant. "Industrial augmented reality: Requirements for an augmented reality maintenance worker support system." In *2018 IEEE International Symposium on Mixed and Augmented Reality Adjunct (ISMAR-Adjunct)*, pp. 151–153. IEEE, 2018.
5. Zhang, Zhenliang, Dongdong Weng, Haiyan Jiang, Yue Liu, and Yongtian Wang. "Inverse augmented reality: A virtual agent's perspective." In *2018 IEEE International Symposium on Mixed and Augmented Reality Adjunct (ISMAR-Adjunct)*, pp. 154–157. IEEE, 2018.
6. Raju, K. Chalapathi, et al. "3D based modern education system using augmented reality." In *2018 IEEE 6th International Conference on MOOCs, Innovation and Technology in Education (MITE)*, pp. 37–42. IEEE, 2018.
7. Karkera, Akshay, Sushil Dhadse, Vinayak Gawde, and Kavita Jain. "Pokémon fight augmented reality game." In *2018 Second International Conference on Inventive Communication and Computational Technologies (ICICCT)*, pp. 1762–1764. IEEE, 2018.
8. Shibata, Yoshitaka, and Katsumi Sasaki. "Tourist information system based on beacon and augmented reality technologies." In *2016 19th International Conference on Network-Based Information Systems (NBiS)*, pp. 410–413. IEEE, 2016.
9. Sato, Goshi, Go Hirakawa, and Yoshitaka Shibata. "Push typed tourist information system based on beacon and augmented reality technologies." In *2017 IEEE 31st International Conference on Advanced Information Networking and Applications (AINA)*, pp. 298–303. IEEE, 2017.
10. Pucihar, Klen Čopič, and Paul Coulton. "Exploring the evolution of mobile augmented reality for future entertainment systems." *Computers in Entertainment (CIE)* 11, no. 2 (2015): 1–16.
11. Agarwal, Charvi, and Narina Thakur. "The evolution and future scope of augmented reality." *International Journal of Computer Science Issues (IJCSI)* 11, no. 6 (2014): 59.
12. Arth, Clemens, Raphael Grasset, Lukas Gruber, Tobias Langlotz, Alessandro Mulloni, and Daniel Wagner. "The history of mobile augmented reality." arXiv preprint arXiv:1505.01319 (2015).

13. Mekni, Mehdi, and Andre Lemieux. "Augmented reality: Applications, challenges and future trends." *Applied Computational Science* (2014): 205–214.

14. Barsom, Esther Z., Maurits Graafland, and Marlies P. Schijven. "Systematic review on the effectiveness of augmented reality applications in medical training." *Surgical Endoscopy* 30, no. 10 (2016): 4174–4183.

15. Zhu, Wei, Charles B. Owen, Hairong Li, and Joo-Hyun Lee. "Personalized in-store e-commerce with the promopad: an augmented reality shopping assistant." *Electronic Journal for E-commerce Tools and Applications* 1, no. 3 (2004): 1–19.

16. Vlahakis, Vassilios, M. Ioannidis, John Karigiannis, Manolis Tsotros, Michael Gounaris, Didier Stricker, Tim Gleue, Patrick Daehne, and Luís Almeida. "Archeoguide: an augmented reality guide for archaeological sites." *IEEE Computer Graphics and Applications* 22, no. 5 (2002): 52–60.

17. Huang, Zhanpeng, Weikai Li, Pan Hui, and Christoph Peylo. "CloudRidAR: A cloud-based architecture for mobile augmented reality." In *Proceedings of the 2014 Workshop on Mobile Augmented Reality and Robotic Technology-Based Systems*, pp. 29–34. 2014.

18. Hsiao, Kuei-Fang. "Using augmented reality for students health-case of combining educational learning with standard fitness." *Multimedia Tools and Applications* 64, no. 2 (2013): 407–421.

19. Hughes, Charles E., Christopher B. Stapleton, Darin E. Hughes, and Eileen M. Smith. "Mixed reality in education, entertainment, and training." *IEEE Computer Graphics and Applications* 25, no. 6 (2005): 24–30.

20. Hwang, Gwo-Jen, Po-Han Wu, Chi-Chang Chen, and Nien-Ting Tu. "Effects of an augmented reality-based educational game on students' learning achievements and attitudes in real-world observations." *Interactive Learning Environments* 24, no. 8 (2016): 1895–1906.

21. Wu, Hsin-Kai, Silvia Wen-Yu Lee, Hsin-Yi Chang, and Jyh-Chong Liang. "Current status, opportunities and challenges of augmented reality in education." *Computers & Education* 62 (2013): 41–49.

22. Masood, Tariq, and Johannes Egger. "Augmented reality in support of Industry 4.0—Implementation challenges and success factors." *Robotics and Computer-Integrated Manufacturing* 58 (2019): 181–195.

23. Hayhurst, Jason. "How augmented reality and virtual reality is being used to support people living with dementia—Design challenges and future directions." In Timothy Jung and M. Claudia tom Dieck (eds.) *Augmented Reality and Virtual Reality*, pp. 295–305. Springer, Cham, 2018.

24. Pyae, Aung, and Leigh Ellen Potter. "A player engagement model for an augmented reality game: A case of Pokémon go." In *Proceedings of the 28th Australian Conference on Computer-Human Interaction*, pp. 11–15. 2016.

25. Roesner, Franziska, Tadayoshi Kohno, and David Molnar. "Security and privacy for augmented reality systems." *Communications of the ACM* 57, no. 4 (2014): 88–96.

26. Ban, Yuki, Takuji Narumi, Tatsuya Fujii, Sho Sakurai, Jun Imura, Tomohiro Tanikawa, and Michitaka Hirose. "Augmented endurance: controlling fatigue while handling objects by affecting weight perception using augmented reality." In *Proceedings of the SIGCHI Conference on Human Factors in Computing Systems*, pp. 69–78. 2013.
27. Liao, Tony. "Future directions for mobile augmented reality research: Understanding relationships between augmented reality users, nonusers, content, devices, and industry." *Mobile Media & Communication* 7, no. 1 (2019): 131–149.
28. Parekh, Pranav, Shireen Patel, Nivedita Patel, and Manan Shah. "Systematic review and meta-analysis of augmented reality in medicine, retail, and games." *Visual Computing for Industry, Biomedicine, and Art* 3, no. 1 (2020): 1–20.
29. Vyas, Daiwat Amit, and Dvijesh Bhatt. "Augmented Reality (AR) applications: A survey on current trends, challenges, & future scope." *International Journal of Advanced Research in Computer Science* 8, no. 5 (2017): 2724–2730.
30. Studen, Laura, and Victor Tiberius. "Social media, Quo Vadis? Prospective development and implications." *Future Internet* 12, no. 9 (2020): 146.

CHAPTER 8

Emerging Technologies and Cyber Security

New Horizons of Cyber Security Solutions

Neha Tyagi, Hoor Fatima, and Nitin Rakesh

Sharda University

CONTENTS

DOI: 10.1201/9781003121466-8

8.1 INTRODUCTION

Cybersafety has never been an easy task to implement. Various malicious attacks are emerging every day which impacts the transaction of data. Cybersecurity protects data from cyber threats while transferring it in-between interconnected systems such as hardware, software, and peripheral devices. Cybersecurity practices prevent unauthorized persons from gaining access to data hubs. The aim of implementing cybersecurity practices is to provide the best security solutions in-between the servers, networks, peripheral devices, etc. Furthermore, they are also implemented to secure data from the malicious attacks of hackers. These cyber-attacks are intended to extract, access, delete, or alter the organizations' data or secure the data from the individual. These practices can be applied in many organizations such as government, corporate, financial, medical, etc.

With the development of technologies for cyber-attacks, the field of cybersecurity is also changing exponentially. Although several security protocols are working in the system, there are still many threats of viruses and phishing which need more security for the transactions. For protecting the organizations, services, individuals from cyber-attacks, various utilities such as firewalls, antivirus software, and other technological solutions for protecting personal data and computer networks are important, but they aren't enough. Since society is rapidly developing its cyber-infrastructure, we must educate our populace about how to use it properly. Starting at a young age, cyber ethics, cyber safety, and cybersecurity topics must be incorporated into the educational process. By preventing significant asset losses from cybersecurity threats, security countermeasures help ensure the confidentiality, availability, and integrity of information systems. Cybersecurity has recently emerged as a well-established discipline for computer systems and networks, with a focus on the safety of sensitive data stored on those systems from adversaries seeking to acquire, corrupt, harm, kill, or prevent access to it.

The area of cybersecurity can be classified into various categories:

- **Network security**: There is a need to secure the communication of the organizational members from intruders.

- **Software security**: The software and the applications need to be secured to prevent the loss of data.

- **User security**: There is a need to secure the user by giving them private keys to avoid cyber threats.

- **Organizational security**: There is a huge quantity of data in every organization that needs to be handled while transporting.

- **Data security**: There is a need to secure the data from intruders and hackers. They hack the data for their personal and financial needs.

A few studies on the security ramifications of AI have been published. In contrast to past overviews and investigations, this section centers around far-reaching dangers and procedures to guard during preparing and testing or construing AI from an information-driven point of view.

Furthermore, this section centers around a definite comprehension of the kinds of different digital dangers and components of online protection such as application security, data security, network security, calamity recuperation/business congruity arranging, operational security, end-client instruction, advantages of online protection, network protection challenges, and new patterns in online protection.

8.2 TYPES OF SECURITY THREAT

A computer virus is a malicious computer program that spreads from one computer to another or from one network to another without the user's knowledge [1]. It has the power to erase archives, delete or harm sensitive records, and format hard drives (Figure 8.1).

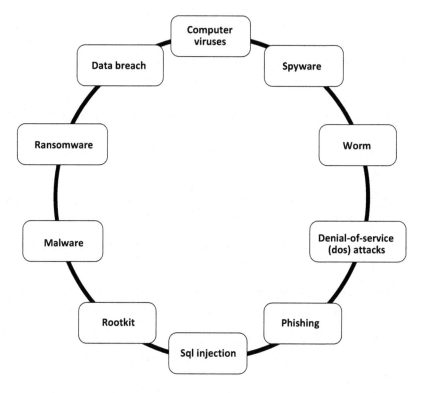

FIGURE 8.1   Types of security threats.

a. What causes a pandemic?

b. Pestilence can be transmitted or attacked in a variety of ways, including:

  i.   Executing an executable.

  ii. Free software and program download and installation.

  iii. Visiting a website that is corrupted and unprotected.

  iv. Clicking on commercials.

  v.   Used tainted USB drives and other removable storage devices.

  vi. Downloading free sports, toolbars, video players, and other software.

  vii. Clicking on a link in a spam email or a URL attachment.

8.2.1 Spyware

Spyware is an undesirable security danger to organizations that introduce itself on the client's gadget and assembles classified data such as individual or business data, login qualifications, and MasterCard data without the client's mindfulness [2]. This kind of danger monitors your online exercises, logs your login certifications, and tracks your subtleties.

Thus, any organization or individual should find ways to forestall spyware by introducing hostile to infection programming, arranging a firewall, and downloading programming from trustworthy sources.

8.2.2 Worm

A PC worm is a kind of malignant programming or program that spreads across an organization and duplicates itself starting with one PC then onto the next inside an organization. It can spread without the assistance of people, abusing programming security blemishes, and endeavoring to take classified data, degenerate information, and introduce a secondary passage for far off admittance to the system [3].

8.2.3 Denial-of-Service (DoS) Attacks

Denial-of-Service (DoS) is an assault that debilitates or delivers a PC or organization out of reach to clients. It typically floods a focused gadget with demands until regular traffic can't be taken care of, bringing about clients for swearing of administration. At the point when an aggressor keeps real clients from getting to explicit PC frameworks, PCs, or different assets [3], this is known as disavowal of the administration assault.

The assailant sends a huge measure of traffic to the objective worker, over-burdening it, and causing sites, email workers, and other web-associated administrations to go down.

8.2.4 Phishing

Phishing is a kind of friendly designing assault that endeavors to get data, for example, usernames, passwords, MasterCard numbers, login qualifications, etc. In a phishing email attack, an interloper sends phishing messages to the casualty's email address that have all the earmarks of being from their bank and request that they give individual information [4]. The message includes a connection that will redirect you to a different insecure website where your information will be stolen. It is, therefore, best to avoid or not click or open such an email and not include your confidential information.

8.2.5 SQL Injection

SQL infusion is a type of infusion assault and quite possibly the most well-known web hacking method that permits the aggressor to alter or eliminate information in the backend database [4]. It's an application security defect, supposing that an application doesn't as expected clean SQL articulations, an aggressor may utilize noxious SQL orders to access the association's data set. The vindictive code is installed in SQL proclamations and goes into an online page by the assailant.

8.2.6 Rootkit

A rootkit might be a PC program that unpretentiously introduces and runs malicious code on a PC or orchestrates a machine without the client's data in organizing to get manager level data to the system [5]. Rootkits, Firmware Rootkits, Kernel-Level Rootkits, and Application Rootkits are the other states of rootkit infections. It is regularly undermined while utilizing a contraption, either by the sharing of corrupted circles or drives. It's usually presented without the casualty's data, either with an undermined secret word or by using coding flaws, social structure methodologies, and phishing techniques.

8.2.7 Malware

Malware is a type of software created by cybercriminals that normally comprises a program or code [5]. It envelops a wide scope of online protection chances pointed toward unleashing ruin on networks or acquiring unapproved admittance to a framework. Malware can likewise ruin an instrument by being shipped off the client as a connection or a document in an email, permitting the client to tap on the connection or open the record to execute the malware. This sort of assault incorporates PC infections, worms, Trojan ponies, and spyware.

8.2.8 Ransomware

Ransomware could be a frame of security risk that anticipates clients from getting to the ADP framework and requests bitcoin in trade to decrypt them. WannaCry, Petya, Cerber, Locky, and CryptoLocker are some of the foremost harming ransomware assaults [6].

8.2.9 Data Breach

An information breach may be a security danger in which delicate or secure data is uncovered and data is obtained from a gadget without the owner's consent. MasterCard numbers, buyer records, exchange of privileged insights, and another individual, exclusive, or private data may be included.

8.3 ELEMENTS OF CYBERSECURITY

8.3.1 Application Security

Jeong's utility of protection in net utility is of profound significance because of the prolonged use of the net for business [7]. Most of the assaults are due to the fact that the builders aren't thinking about protection as a challenge or because of the safety flaws in designing and growing the applications. The enforcement of protection inside the software program improvement lifestyle cycle of the utility can also additionally lessen the excessive value and efforts related to imposing protection at a later stage. For this purpose, diverse tries have been made to outline a few protection styles maintaining the assaults in mind. The builders now can use those styles; however, from time to time it's far tough to select a sample from the huge list, which can also add or might not relate to the context. Due to the quantum bounce in functionality, the fee of upgrading conventional cellular telephones to smartphones is tremendous. One of the maximum appealing functions of smartphones is the provision of a huge variety of apps for customers to download and install. However, in additional way, hackers can without problems distribute malware to smartphones, launching diverse attacks. This difficulty needs to be addressed with the aid of using each preventive process and powerful detection techniques. While clients are energetic on downloading applications from application showcases, this gives programmers a helpful way to pollute cell phones with malware. For instance, they could repackage celebrated computer games with malware and disperse them inside the application markets. As often as possible clients are keen on downloading the aroused applications. The most recent study articulated that 267,259 malware-aroused applications were found, among which 254,158 are live on the Android stage.

Because of the widespread use of the internet for business, the application of security in web applications is critical. The majority of attacks occur as a result of developers failing to prioritize protection or as a result of safety laws in the planning and development of apps. The application's social control of security during the software package creation life cycle

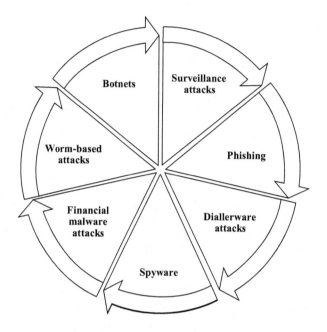

FIGURE 8.2   Elements of cybersecurity.

can reduce the high cost and commitment of enforcing security at a later level. Different tests have been produced for this reason to lay out certain security patterns while holding assaults at the top of the priority list. The engineers presently will utilize these examples anyway, normally it's irksome to pick an example from the enormous rundown, which could suit the unique situation (Figure 8.2).

8.3.2 Information Security

Security failures in the early days of computing were predominantly viruses and worms that would flash a message or a message on the computer advertising without triggering any significant harm to the data or devices that are being used [8]. Furthermore, rare instances of attacks with the ability to damage data were reported. There was, for example, the Friday the 13th virus, which was intended to delete any program run that day. On a certain Friday, all the data on the contaminated disc drive is deleted. Information technology has grown from handling small and harmless violations of technology to managing those with a major effect on the economic development of organizations. The security of data isn't tied in with taking a gander at the past with irateness at an attack once confronted; nor is it about taking a gander at the present with dread of being assaulted; nor is it about viewing at the future in disarray concerning what could befall us. The message is that, consistently, associations and individuals should be ready.

The assaults against such a computing are genuine and physical. It very well might be important to avoid potential risks to decrease them to a worthy sum. Both PC programs were overseen securely [9]. Electronically, it has been difficult for any gathering to get unapproved entry to information. There had been no security threat reported by the customers, also no approaches or any subtleties electronically. Tragically (from a guard office, viewpoint), gadget use has developed. Also, these security insurances have become obsolete, and further security steps that are required have been presented.

8.3.3 Network Security

With the exponential growth of the network and communication infrastructure, there has been a growing range of network and communication technologies, focusing on the integrity of networks. Networking data generated, created, or extracted from the network system typically reflects protection via analyzing the data relating to network security incidents and network security activities. You will calculate and measure the network device we are referencing. Data that suggests security risks and demonstrates infrastructure, security, privacy, and trust anomalies as well as confidentiality. Network data relates to defense, in short, security-related data. Gathering cyber-related data security hazards and economic damages incurred by network assaults, intrusions, and bugs is the first step in detecting network attacks and intrusions. An intensive study on network security has been promoted. Anomalies in traffic are characterized by rare and major anomalies. Alterations in the behavior of the network traffic are also observed.

8.3.4 Business Continuity Planning

Essential service agencies perform vital functions in the preservation of public health after a disaster; thus, preparation for business continuity is essential to ensuring that key activities continue to operate after a crisis. Typically, corporate continuity planning is based on a method to anticipate and stop. Asset mapping exercises have the potential to complement the primarily risk-based approach by reflecting on corporate resources and skills. In March and April 2014, two focus group meetings with members of vital support organizations (n=22) were conducted in Ottawa, Canada, using the facilitation format of the Standardized Interview Matrix. Eight emerging trends that illustrate the value of organizational-level assets and their contribution to adaptive ability were defined using inductive reasoning [10].

8.3.5 End-User Education

Kudjo, the emergence of the age of computers, has inspired a growing interest in the simple data sets that can now easily be accessed, stored, and are under investigation. Security has many facets and several of them are applications, spanning from stable trading and transfers to transfer secret contact and password protection. Also, the multiple threats posed by cyber-attacks cause computer education and technology literacy criteria: Confidentiality. Any organization's key knowledge and knowledge is thus wise and often tested and used by end-users. To teach users about the existing security systems that can be used and implemented to secure these properties efficiently and effectively. The latest cyber-attacks on the country's institutions identify the need for education and understanding of informative prevention concerns [11].

8.4 CYBERSECURITY CHALLENGES

Cybersecurity is an important factor behind the country's extensive political and economic security plans. In a developing nation like India, there are several concerns about cybersecurity. Because of the increase in cyber-attacks, there is a vital need for a security expert in organizations to ensure that their networks are secure and reliable [12]. These technology experts deal with a wide range of cybersecurity issues, including securing private company servers, shielding classified data held by presidential organizations, and so on (Figure 8.3).

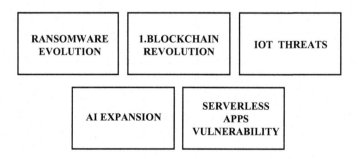

FIGURE 8.3   Various challenges in cybersecurity.

8.4.1 Ransomware Evolution

Ransomware is a form of malware that encrypts data on a target's computer and demands compensation before the data is decrypted. After a satisfactory payment, the victim's access rights are returned. Ransomware is the curse of cyber defense, data analysts, IT, and executives. Ransomware threats are on the rise in the world of cybercrime. IT experts and corporate executives should have a solid response strategy in place to shield their enterprise from ransomware attacks. It involves following the correct procedures for recovering corporate and customer data and software, as well as reporting all data breaches in compliance with the Notifiable Data Breach scheme. DRaaS tools offer the most effective protection against ransomware attacks today. Using the DRaaS solutions app, we can automatically replicate our files, quickly recognize that the backup is safe, and launch a fail-over with the push of a button when malicious attacks infect our records.

8.4.2 Blockchain Revolution

Blockchain innovation is a foremost noteworthy innovation within the computing period. It is the essential time in human history that we take a local digital intermediate for peer-to-peer esteem trade. The blockchain may well be an innovation that grants cryptocurrencies like Bitcoin. The blockchain may be an endless worldwide stage that empowers two or more parties to undertake and do an exchange or do commerce without having a third party for setting up belief [13]. It is troublesome to anticipate what blockchain frameworks will offer concerning cybersecurity. The cybersecurity experts can make a few surmises concerning blockchain. Since the application and usefulness of the blockchain in a cybersecurity setting develop, there'll be solid pressure but too complementary integrative with conventional demonstrated cybersecurity approaches.

8.4.3 Internet of Things (IoT) Threats

The Internet of Things (IoT) is a term that refers to a network of computers that are linked to the internet. It's a network of web-connected physical devices that can be downloaded over the internet. The physical devices attached to the network have a unique identifier (UID) and can transmit and receive data over a network without involving human-to-human or human-to-computer interactions [14]. Consumers and companies were vulnerable to cyber-attacks because of the firmware and programming programs that run on IoT computers. It was not anticipated when IoT devices were first launched that they would be used for cybersecurity and commercial purposes. Each corporation should collaborate with cyber protection experts to verify the security of their authentication policies, session management, customer validation, multifactor confirmation, and security conventions to better handle the risk.

8.4.4 Serverless Apps Vulnerability

Applications that depend on the outsider cloud framework or back-end administrations such as Google Cloud Functions, Amazon Web Services (AWS) Lambda, and other comparable administrations are known as serverless engineering and applications. Fox is a developed character. Serverless applications welcome digital aggressors to immediately spread assaults on their machines when clients access the apparatus locally or off-worker on their PCs. As an outcome, the serverless application's safety efforts are the obligation of the client. The serverless applications never really keep aggressors from getting to our information. The serverless application would not be of assistance if an aggressor accesses our information through a secondary passage like released passwords, an undermined insider, or hacking. We can run it with instruments, giving us the most obvious opportunity with regard to foiling cybercriminals. Applications that needn't bother a worker are generally downsized. It empowers engineers to dispatch their applications rapidly and without any problem. They ought not to be stressed over the foundation. Famous workers with fewer applications incorporate web administrations and preparing instruments.

8.5 CYBERSECURITY RISK ANALYSIS

- **Conduct a risk assessment survey**: Getting management and department heads' feedback is crucial to the risk assessment process. The danger assessment survey refers to the process of identifying and reporting the specific risks or threats that each department faces.

- **Determine the dangers**: This phase is used to assess an IT system or other aspects of a business to identify the risks associated with the software, hardware, data, and IT personnel. It identifies the potential for negative events in a business, such as human error, flooding, fire, or earthquakes.

- **Analyze the risks**: Once the risks have been assessed and identified, the danger analysis process should examine each potential risk as well as the consequences associated with each risk. It also decides how they can impact an IT project's goals.

- **Develop a risk management strategy**: After analyzing the risk and considering which assets are important and which risks are likely to negatively impact IT assets, we will create a risk management strategy to include control advice that would be used to reduce, move, embrace, or avoid the risk.

- **Implement the change management plan**: The primary aim of this phase is to put in place steps to eliminate or minimize the risks associated with the study. We can eliminate or reduce the threat by dealing with the highest priority and resolving or mitigating each risk to the point that it is no longer a threat.

- **Monitor the risks**: This phase is in charge of tracking the security risk regularly to detect, handle, and manage risks, which should be an integral part of every risk analysis process [15].

8.6 TRENDS IN CYBERSECURITY

In this segment, we examine a few utilizations of AI in network protection. Different areas such as establishment security, mechanical control frameworks, interruption discovery in administrative control and information obtaining (SCADA) frameworks, interruption recognition for a vehicular impromptu organization (VANET), malware examination, and so on have seen the use of the AI procedures as of late [16]. During this short audit, we don't intend to be extensive. We test a couple of guides to supply the perusers with a sample of the relevance of AI in network safety.

- **Network security**: Network security leads to identify malicious applications over the network for providing the confidentiality and integrity of the data over the network. Many systems have been designed based on machine learning approaches to detect intruders over the network. There will be an improved algorithm for Genetic Network Programming (GNP) which is capable of the detection of misuse and anomaly [17]. The unsupervised learning algorithm has been proposed over the KDD cup 99 data for intrusion detection. The result shows the accuracy of a 94.91% false-positive rate. A ready characterization framework utilizing Neural Networks and SVMs have been created to distinguish the Distributed DoS assaults. The proposed approach has 95% exactness with the individual precision of a neural organization as 83% for the Neural Network and 99% exactness with SVM [18]. A crossover bunching procedure and the SVM have been utilized to recognize the interruption in a remote sensor organization. The proposed approach shows a precision of 98.47 with seven highlights and a least of 96.95% with seven highlights [19].

- **Software security**: It is imperative to secure IT resources like applications. Moreover, the new arrangement of executing a profound cautious technique isn't adequate to get IT resources as it solidifies the layer of the organization. However, it is not the layer of the application where most programming works [20]. Organizations need programming to buy direction, designers need guidance, what to assess, and programming support organizers need to know. An assessment of things to come should be accurately anticipated programming issues that they can experience [21]. The requirement for the advancement of safe programming is difficult that influences all PC business partners. Bugs in applications are one of the lasting issues in the improvement of applications. Not simply absconds, it does, nonetheless, lead to a penetration of safety. The vast majority of the shortcomings are not identified with security concerns, like blunders in code, execution, and issues with convenience. Notwithstanding, past analysts have brought up that some product bugs can make programming helpless against assaults, so we can group these bugs as security-related [22].

- **User security**: The comparison of techniques for detecting phishing occurs for ensuring the user's security [23]. It has been noted that there is a high rate of missed detection in many phishing detection solutions under consideration. Utilizing 1,171 crude phishing messages and 1,718 authentic messages, specialists looked at six machines learning classifiers: "Strategic relapse (LR), Classification and Regression Trees (CART), Bayesian Additive Regression Trees (BART), SVM, Random Forests (RF), and Neural Networks (NNets)."

- A robotized structure for phishing locations was created by Zhuang et al. [24], utilizing a group outfit of a few bunching arrangements. An element determination calculation was utilized to remove diverse phishing email attributes, which was a cosine likeness (utilizing the TF IDF metric) progressive bunching (HC) calculation for ascertaining the comparability between two focuses and a K-Medoids (KM) grouping approach. The proposed web phishing and malware order methods have around 85% proficiency.

- **Organizational security**: In general, organizational learning involves the following elements, such as qualified individuals, organizational expertise and experience collection, individual independence, and infrastructure creation for exchange between stakeholders, experts, and stored experience [25].

- **Data security**: This segment centers around the security and privacy of the actual information. As the world moves into the Big Data period, present-day classifier models (particularly DNNs) require a lot of information for preparing. Since publicly supporting has become a mainstream technique for social affair information, it represents a critical danger of releasing touchy information like photographs, recordings, individual data, clinical records, etc.

- Erlingsson et al. [26] proposed RAPPOR, a mysterious and hearty publicly supporting system that joined randomized reaction with DP to guarantee publicly supporting security. Moreover, specialists have as of late utilized DP to ensure the protection of an assortment of learning calculations, including SVM [Benjamin], profound learning [27], and Bayesian improvement [28]. Besides, homomorphic encryption (HE) [29] is another strategy for guaranteeing information security through encryption. Numerous scientists host investigated safe multi-get-together computations [30,31], full HE information arrangement [32], circulated k-implies grouping calculations, and encoded information taking care of neural organizations dependent on HE [33].

8.7 CONCLUSIONS AND FUTURE WORK

Machine learning is widely used in a variety of practical applications, including image processing, natural language processing, pattern recognition, computer vision, intrusion detection, malware identification, autonomous driving, and security defense.

Machine learning is becoming increasingly important in both training and inference processes. A comprehensive survey of security problems using a variety of machine learning techniques is presented in this chapter. Machine learning poses security threats in two areas: the training process and the testing/inferring process. Vulnerability assessment frameworks, countermeasures in the training process, those in the testing or inference process, data protection, and privacy are all examples of current protective machine learning techniques.

In general, this chapter is meant to address network safety information science and related methodologies as well as the materialness of online protection frameworks and administrations to information-driven astute dynamic from an AI viewpoint. Our investigation and conversation may have a wide scope of suggestions for both security specialists and experts. We have laid out a few issues and exploration headings for analysts to consider.

REFERENCES

1. A. Alnasser, H. Sun, and J. Jiang, "Cybersecurity challenges and solutions for V2X communications: A survey," *Comput. Networks*, vol. 151, pp. 52–67, 2019, doi: 10.1016/j.comnet.2018.12.018.
2. Robinson et al, "City research online city, university of London institutional repository," *City Res. Online*, vol. 37, no. 9, pp. 1591–1601, 2008.
3. M. Omar, "A world of cyber attacks (a survey)," 2019.
4. S. N. Hussain, "A survey on cyber security threats and their solutions," *Int. J. Res. Appl. Sci. Eng. Technol.*, vol. 8, no. 7, pp. 1141–1146, 2020, doi: 10.22214/ijraset.2020.30449.
5. A. Handa, A. Sharma, and S. K. Shukla, "Machine learning in cybersecurity: A review," *Wiley Interdiscip. Rev. Data Min. Knowl. Discov.*, vol. 9, no. 4, pp. 1–7, 2019, doi: 10.1002/widm.1306.
6. I. H. Sarker, A. S. M. Kayes, S. Badsha, H. Alqahtani, P. Watters, and A. Ng, "Cybersecurity data science: An overview from a machine learning perspective," *J. Big Data*, vol. 7, no. 1, 2020, doi: 10.1186/s40537-020-00318-5.
7. C. Y. Jeong, S. Y. T. Lee, and J. H. Lim, "Information security breaches and IT security investments: Impacts on competitors," *Inf. Manag.*, vol. 56, no. 5, pp. 681–695, 2019, doi: 10.1016/j.im.2018.11.003.
8. A. K. Dalai and S. K. Jena, "Evaluation of web application security risks and secure design patterns," *ACM Int. Conf. Proc. Ser.*, no. January, pp. 565–568, 2011, doi: 10.1145/1947940.1948057.
9. M. T. Dlamini, J. H. P. Eloff, and M. M. Eloff, "Information security: The moving target," *Comput. Secur.*, vol. 28, no. 3–4, pp. 189–198, 2009, doi: 10.1016/j.cose.2008.11.007.

10. A. Salam, *Internet of things for sustainability: Perspectives in privacy, cyber-security, and future trends.* 2020.

11. M. E. Thomson and R. Von Solms, "Information security awareness: Educating your users effectively," *Inf. Manag. Comput. Secur.*, vol. 6, no. 4, pp. 167–173, 1998, doi: 10.1108/09685229810227649.

12. B. Hamid, N. Jhanjhi, M. Humayun, A. Khan, and A. Alsayat, "Cyber security issues and challenges for smart cities: A survey," *MACS 2019-13th Int. Conf. Math. Actuar. Sci. Comput. Sci. Stat. Proc.*, pp. 1–7, 2019, doi: 10.1109/MACS48846.2019.9024768.

13. F. Jameel, U. Javaid, W. U. Khan, M. N. Aman, H. Pervaiz, and R. Jäntti, "Reinforcement learning in blockchain-enabled IIoT networks: A survey of recent advances and open challenges," *Sustainability*, vol. 12, no. 12, pp. 1–22, 2020, doi: 10.3390/su12125161.

14. V. Hassija, V. Chamola, V. Saxena, D. Jain, P. Goyal, and B. Sikdar, "A survey on IoT security: Application areas, security threats, and solution architectures," *IEEE Access*, vol. 7, pp. 82721–82743, 2019, doi: 10.1109/ACCESS.2019.2924045.

15. Y. Cherdantseva et al., "A review of cybersecurity risk assessment methods for SCADA systems," *Comput. Secur.*, vol. 56, pp. 1–27, 2016, doi: 10.1016/j.cose.2015.09.009.

16. R. Mitchell and I. R. Chen, "A survey of intrusion detection techniques for cyber-physical systems," *ACM Comput. Surv.*, vol. 46, no. 4, 2014, doi: 10.1145/2542049.

17. N. Lu, S. Mabu, T. Wang, and K. Hirasawa, "An efficient class association rule-pruning method for unified intrusion detection system using genetic algorithm," *IEEJ Trans. Electr. Electron. Eng.*, vol. 8, no. 2, pp. 164–172, 2013, doi: 10.1002/tee.21836.

18. P. S. Saini, S. Behal, and S. Bhatia, "Detection of DDoS attacks using machine learning algorithms," *Proc. 7th Int. Conf. Comput. Sustain. Glob. Dev. INDIACom 2020*, no. March, pp. 16–21, 2020, doi: 10.23919/INDIACom49435.2020.9083716.

19. H. Sedjelmaci and M. Feham, "Novel hybrid intrusion detection system for clustered wireless sensor network," *Int. J. Netw. Secur. Its Appl.*, vol. 3, no. 4, pp. 1–14, 2011, doi: 10.5121/ijnsa.2011.3401.

20. Labbé, A. (2018). PRIMER: Banks and cyber security (part 1). International Financial Law Review, Retrieved from https://www.proquest.com/scholarly-journals/primer-banks-cyber-security-part-1/docview/2012825779/se-2?accountid=133553

21. O. H. Alhazmi, Y. K. Malaiya, and I. Ray, "Measuring, analyzing and predicting security vulnerabilities in software systems," *Comput. Secur.*, vol. 26, no. 3, pp. 219–228, 2007, doi: 10.1016/j.cose.2006.10.002.

22. F. Camilo, A. Meneely, and M. Nagappan, "Do bugs foreshadow vulnerabilities? A study of the chromium project," *IEEE Int. Work. Conf. Min. Softw. Repos.*, vol. 2015-Augus, pp. 269–279, 2015, doi: 10.1109/MSR.2015.32.

23. S. Abu-nimeh, D. Nappa, X. Wang, and S. Nair, "P60-Abu-Nimeh.Pdf," pp. 60–69, 2007.
24. W. Zhuang, Y. Ye, Y. Chen, and T. Li, "Ensemble clustering for internet security applications," *IEEE Trans. Syst. Man Cybern. Part C Appl. Rev.*, vol. 42, no. 6, pp. 1784–1796, 2012, doi: 10.1109/TSMCC.2012.2222025.
25. Y. Lu and L. Da Xu, "Internet of things (IoT) cybersecurity research: A review of current research topics," *IEEE Internet Things J.*, vol. 6, no. 2, pp. 2103–2115, 2019, doi: 10.1109/JIOT.2018.2869847.
26. Ú. Erlingsson, V. Pihur, and A. Korolova, "Rappor," pp. 1054–1067, 2014, doi: 10.1145/2660267.2660348.
27. M. Abadi et al., "Deep learning with differential privacy," *Proc. ACM Conf. Comput. Commun. Secur.*, vol. 24–28-Octo, no. Ccs, pp. 308–318, 2016, doi: 10.1145/2976749.2978318.
28. M. J. Kusner, J. R. Gardner, R. Garnett, and K. Q. Weinberger, "Differentially private Bayesian optimization," *32nd Int. Conf. Mach. Learn. ICML 2015*, vol. 2, pp. 918–927, 2015.
29. W. Hu and Y. Tan, "Generating adversarial malware examples for black-box attacks based on GAN," 2017, [Online]. Available: http://arxiv.org/abs/1702.05983.
30. O. A. Alimi et al., "Applications of machine learning in cybersecurity," *Comput. Secur.*, vol. 8, no. 1, pp. 1–7, 2019, doi: 10.22214/ijraset.2020.30449.
31. I. Damgård, V. Pastro, N. Smart, and S. Zakarias, "Multiparty computation from somewhat homomorphic encryption," *Lect. Notes Comput. Sci. (including Subser. Lect. Notes Artif. Intell. Lect. Notes Bioinformatics)*, vol. 7417 LNCS, pp. 643–662, 2012, doi: 10.1007/978-3-642-32009-5_38.
32. L. J. M. Aslett, P. M. Esperança, and C. C. Holmes, "Encrypted statistical machine learning: New privacy-preserving methods," pp. 1–39, 2015, [Online]. Available: http://arxiv.org/abs/1508.06845.
33. N. Dowlin et al., "CryptoNets: Applying neural networks to encrypted data with high throughput and accuracy - Microsoft research," *Microsoft Res. TechReport*, vol. 48, pp. 1–12, 2016, [Online]. Available: http://proceedings.mlr.press/v48/gilad-bachrach16.pdf%0Ahttp://research.microsoft.com/apps/pubs/?id=260989.

CHAPTER 9

3D Printing Procedures

Explore the Endless Possibilities of 3D Object Design, Modeling and Manufacturing

Rajesh Pant, Pankaj Negi, and Jasmeet Kalra

Graphic Era Hill University

Sandeep Tiwari

Krishna Engineering College

CONTENTS

DOI: 10.1201/9781003121466-9

9.1 INTRODUCTION

The 3D printing technique uses a CAD model to create a physical object, usually by stacking various continuous thin layers of material. It converts a CAD object into its physical form by adding materials layer by layer (Figure 9.1).

Various techniques are used to print an object in 3D. It produces two historical innovations: the conversion of objects in their digital format and the manufacture of an object by adding materials. The concept of 3D printing is the addition of materials. As 3D printing is usually another production method based on high-tech where components are produced

FIGURE 9.1 3D printing.

additively in layers at the micron level, this is generally different from other existing conventional production methods.

There are many limitations to conventional production, which is largely manpower-based and handmade. An ideology derived from the etymological roots of the French word for production itself. However, the production era has new and automated techniques such as machining, drilling, cutting and forming, which are complex manufacturing processes that need machinery, computers, and robust techniques.

9.2 3D PRINTING TECHNOLOGY

The first step in this technique is to create a model using CAD. Simpler, more accessible programs are available for the Producer and Consumers – or imaged with a 3D scanner. The model is then "split" into different layers, thus changing the design to a 3D printer compatible file. According to the design material, the 3D will be applied to the 3D printer. As mentioned, there are different 3D printing processes where different materials were used to produce the final part / product in different ways. The daily used polymers, metals are now widely used for research and development purposes. Research is also carried out on biomaterials for 3D printing and various types of food materials. Although at the beginning, the materials were limited. Only plastic is used much, mostly ABS or PLA, but nylon is also an alternative. Different types of new machineries are also developed that have been used to produce foods such as grains and sugars (Figure 9.2).

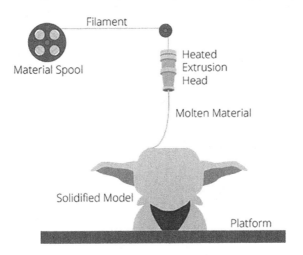

FIGURE 9.2 Example of 3D printing (FDM).

9.3 CLASSIFICATION OF 3D PRINTING

Various types of 3D printing methods are available. Based on the 2D layer formation, 3D printing can be categorized into the following types as shown in Figure 9.3.

   a. **Photopolymerization**

   b. **Extrusion based**

   c. **Powder based**

   d. **Lamination**

9.3.1 Photopolymerization

Photopolymerization-based 3D printing techniques use photosensitive polyresins which are cured in ordered layer using laser or digital light projection method. These photopolymers are made up of photo initiator,

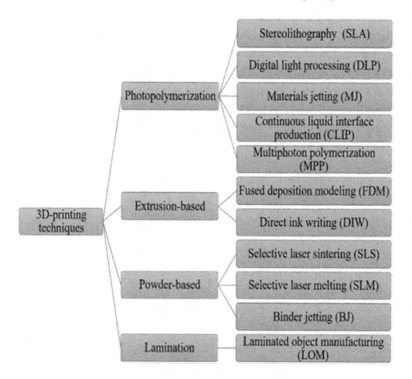

FIGURE 9.3 Classification of 3D printing techniques. (From Ambrosi, A. & Pumera, M., *Chemical Society Reviews*, 45, pp. 2740–2755, 2016. With permission.) [1].

additives and reactive monomers. Few techniques which represent photo-polymerization are Stereolithography (SLA), digital light processing (DLP), materials jetting (MJ), Continuous liquid interface production (CLIP) and multi photon polymerization (MPP). SLA is an additive process in which a light source is utilized to cure a thermoset resin into hardened plastic. It works on the principle that when the thermoset resins are exposed to a fixed wavelength of light, the resins form a polymer chain of monomer or oligomer into a rigid structure. The SLA machine begins to draw structure from exporting a CAD/CAM drawing forming its support structure by focusing ultraviolet laser beam on the surface layer of thermoset resins and thereafter recoating the next layer. This process continues until the complete structure is built. After that post curing is required to clear off any extra resins. Digital light processing (DLP) is like SLA as both use photopolymer. The major difference lies in the use of light source; DLP requires a more conventional light source, like an arc lamp which can directly be applied to the whole surface of photopolymer resins making it faster than SLA. DLP requires a shallow layer of resins which reduces the waste and decreases running cost. Also, DLP produces highly accurate and high-resolution structures in comparison to SLA. However, it also requires the same support system and same post curing. MJ produces parts using liquid photopolymer droplets and curing takes place through UV light as in SLA. MJ is compared with the 2D inkjet process as the process for both is almost same except that MJ deposits layer by layer until the part is finished whereas inkjet deposits a single layer only. MJ provides smooth surfaces and high dimensional accuracy. It utilizes many types of material such as polymers, composites, ceramics, biological and hybrids [2,3]. CLIP uses the oxygen permeable membrane that enhances its speed by a high degree. In CLIP, through an oxygen permeable, continuous series of UV images are projected, which are generated by DLP below a liquid resin bath. The dead zone or area of uncured liquid resin maintains a liquid interface below the part. Polymerization takes place once the oxygen is depleted [4]. MPP is one technique which enables it to cure the photopolymer at any point of the polymer bath. It can receive 100 nm spatial resolution with the help of fast laser.

9.3.2 Extrusion Based

The next type of classification is extrusion based which is represented by fused deposition modeling (FDM) and direct ink writing (DIW).

Extrusion depending on the 3D printing method is common due its simplicity and machine availability. In this technique a semi-solid or liquid state material is directed through a controlled nozzle and is deposited over the platform as per the design. FDM makes 3D objects with extrusion of heated thermoplastic polymers in wires through a nozzle tip to deposit layers over the platform while direct ink writing (DIW) utilizes a liquid phase viscous ink which is dispensed out of nozzles with a controlled flow rate and depositing along the digitally defined contours to build a 3D structure. Ceramics, plastics, food and living cells [5] are some materials which can be used as ink in DIW. There are many advantages like simplicity, versatility, reliability, and affordability of extrusion based techniques. However it also has some weaknesses like low resolution, undesirable characteristics and required print speed for production purpose.

9.3.3 Powder Based

The third classification is powder based which is used on a large scale in industries as it is a basic technique for metal 3D printing. This technique uses the powdered size particles (50–100 µm size) and with the use of laser they are heated and fused together to form a 3D structure. The techniques which fall in this group are selective laser sintering (SLS), selective laser melting (SLM) and binder jetting (BJ). The object built by SLS and SLM are made with a powder material such as nylon which is spread over the build platform in a thin layer. A computer-controlled laser traces the parameters of the object to be built onto the powder. Thereafter it heats the powder either below its melting point (known as 'sintering') or above its melting point (known as 'melting), fusing the particles of powder in a solid form. This is the difference between SLS and SLM - one heats the powder just below the melting point and the other heats it above melting point due to which there is no need for a binding agent in SLM which gives higher compactness and good overall properties rather than SLS. Another way is binder jetting (BJ) which does not use heat during shape formation for metal parts. It deposits the semi-solid glue over a bed of powder which depends on the capillary force to join the powdered particle together. This technology makes use of a jet chemical binder onto the dispersed powder to form the layer [6]. This makes BJ much faster in comparison to SLS and SLM but due to low binding energy between powder and glue, the mechanical properties get affected.

9.3.4 Lamination Based

The last technique of 3D printing is laminated object manufacturing (LOM). A LOM uses rolls of sheet material glued with adhesive to be drawn onto the build platform. A heated roller is used for pressing and melting the adhesive between the sheet material over the build area which undergoes a laser cutting process to produce a 2D layer profile. After each cutting, hot pressing is done to increase the binding. The advantages of LOM are that it can do full-color prints, it is cheap in comparison and is a simpler material handling technique. LOM can produce complicated geometrical parts.

9.4 3D PRINTING MATERIALS

3D printing requires good quality material that meets the specification to produce high-quality structures. To ensure the quality and specification of the product a good control over the procedure, requirement and procurement must be exercised. 3D technology builds a variety of products which are fully functional using raw material as polymers, ceramics, metallic powders, graphene, composites, and their combinations including functionally graded material (FMGs) [2].

9.4.1 Polymers

3D printing technique uses a wide variety of polymers for production of prototypes and working structures with complex geometry [7]. 3D printing uses layer by layer formation of structures using thermoplastic filaments such as acrylonitrile butadiene styrene (ABS), high impact polystyrene [HIPS], polylactic acid (PLA), polypropylene (PP) or polyethylene (PE) [7]. These polymers are widely used in liquid, semi-solid and powdered forms because of their low cost, light weight and high degree of flexibility [8]. Polymer devices are used as inserts in biomechanical devices and healthcare products.

9.4.2 Metals

Many metals which are used in 3D printing technologies are different alloys and stainless steels [9–12]. These metals are used in 3D printing for medical, automobile, aerospace and many other industries. Due to the good physical and chemical properties of metal, they receive high functionality in these industries. Due to the high stiffness, least reactive nature, better resilience and heat-treated conditions, cobalt based alloys are used as dental implants [10]. Due to the high corrosion resistance and high temperature susceptibility of nickel-based alloys, it is used in aerospace and high temperature applications [13]. Titanium alloys have good thermal and mechanical properties which make it easier to use as inserts in the human body and in biomedical applications [12].

9.4.3 Composites

Composites are made by combining two or more materials with different physical and chemical properties so that a new material forms having better properties than its constituents' elements. Due to their varieties, light weight nature and tailored properties they are making significant changes in industries. The principal types are particle reinforced, fiber reinforced and structural composites. The carbon fiber composites are used in automobile and aerospace industries due to their low weight, high strength, high stiffness, low thermal coefficient, good corrosion resistance and better fatigue strength [14]. Similarly, glass fiber reinforced composites are used widely in 3D printing applications having fine properties like high flexibility, resistance to chemical stiffness and high corrosion resistance. Moreover, fiberglass does not burn or get affected by curing temperature.

9.4.4 Ceramics

Ceramics are used traditionally in pottery and kitchenware. But with 3D printing it has revolutionized the industry. 3D printing with ceramics enhances the mechanical properties, product life and quality by eliminating the crack formation and large pores generation in the product [15]. Ceramic products are rigid, durable and fire resistant due to which they can be applied to complex geometries. Ceramics are also used in medical industries as dental implants and in aerospace applications [16]. Some examples of ceramics are alumina, bioactive glasses and zirconia [17,18]. Alumina is an aluminum oxide which is used in a wide range of application including chemicals, aerospace industry, catalyst, adsorbents, microelectronics and advanced-technology industry [19]. SLA machines use ceramics to produce high density and homogenous structures and zirconia is widely used in the construction of nuclear power industries as it has low susceptibility for radiations and low thermal neutron absorption [18].

9.4.5 Smart Materials

Those materials which react to the external stimulus like heat, light, pressure, chemicals and alter their shape or behavior are called smart materials [19]. The smart materials can self-assemble or evolve or completely transform their shapes under external stimuli. These are used in advanced 4 D printing technology. Some of the examples of smart materials are shape memory polymers, shape memory alloys, smart nanocomposites and actuators for soft robotics [20]. These materials can be used in healthcare systems and electromechanical devices. Shape memory polymer can be used to produce any shape using 3D printing technology.

9.4.6 Special Materials

Food items, such as chocolates, candies, pizza, and sauces, can be used in 3D printing to produce desired shape and geometries. 3D food printing can make desired food items without decreasing their nutrient value and taste, making them healthier [21].

9.5 APPLICATION OF 3D PRINTING

9.5.1 Automotive Industry

The automobile industry changes rapidly with time as the demand for better technology and innovation in design, development and production within the shortest possible time is the need of the hour. 3D printing has made phenomenal changes aligning itself with this current need of industry. Ford was one of the earliest companies to adopt 3D printing and now has integrated 3D technology into its product life cycle. The upcoming Shelby GT500 is an example for this as it has 3D printed brake components which have passed all the standards of quality. These parts were built using carbon digital light synthesis technology (DLS) [22]. Volkswagen is a company which after successful test runs is producing all its tooling using 3D printing which reduced its tool production cost by 90% and lead time to minimum. Local motors of Arizona have built the first 3D-printed electric car 'Strati' in 2014. AUDI in collaboration with SLM Solution Group AG produces spare parts and prototypes [23]. 3D technology favors automobile manufacturers in making a prototype, testing, altering and improving the design simultaneously. At the same time 3D technology saves time, money, labor as well as wastage of material to a great extent.

9.5.2 Aerospace Industry

3D printing uses 3D CAD files to print any intricate and complex geometry. The CAD design can be modified and alteration can be saved and printed without wasting time. This became a boon for aerospace industries where earlier the parts were mostly casted taking long time and high investment. 3D gave high flexibility in designing and building parts. 3D printing is widely used in making various engine parts as well as spare parts can be easily made once the drawing is available for replacement or design modification purpose. Mostly nickel based alloys are utilized for aerospace to its high thermal stability, anti-corrosion property and tensile strength.

9.5.3 Food Industry

Food industry is revolutionizing with innovation and production through creativity, nutrient value and sustainability by using food 3D printing. The specialized dietary needs are the demands of children, pregnant women, athletes, actors, health conscious people and patients. The 3D food printing technology is an innovative way to customize food products by additive layer manufacturing derived from CAD software data. The exact amount of mixing of nutrients and giving complex shapes in a diet is the new idea which surfaced in food industry [24]. This can easily be achieved through 3D printing. A nursing home in Germany uses a 3D printing of food called "smooth foods" which is made up of mashed peas, broccoli and carrot, thickened with edible glue to serve elderly people who face difficulty in chewing. The main advantages of using 3D printing in food are that it saves time and effort, is innovative, increases food sustainability and allows to customize the nutrients as per the need. Many edible items like chocolate, pizza, pasta, sauces can be used as raw material to prepare food as per the nutrient diet, hence reducing the negative effects of over accumulation of nutrient and adjust once diet to healthier side [25].

9.5.4 Bio-medical Industry

It is most broadly used in medical industries. It is used by drug and pharmaceutical [26], bone and cartilage replacement [27], tissues regeneration and testing medicines over it, bioprinting of various organs, printing of skin [26], surgical tools and guides, various types of implants, and actual models of patient-specific anatomy and physical structures which allows a medical practitioner to investigate the subject through different angles and come to a medical solution. One can harness many benefits out of 3D printing as it allows practitioner to design and create skin, bones, organs and tissues through it which can be further investigated and researched for drug testing, cosmetics and chemical products.

Pharmaceutical industries can print drugs with accuracy, reproducibility, short time, least cost a complex drug release profile. Prosthetic limbs can be built through this technology that are customized as per the wearer. Practically all the surgical tools and most of the instruments can be produced through 3D printing which decreases the cost and lead time for procurement to minimum.

9.5.5 Architecture and Construction Industry

This is a new vertical in which 3D printing is rapidly developing. An Italian architectural company CLS Architect, along with a UK design and engineering firm Arup construction, built a 100 square meter single story house in one week. The time and wastage are minimum as well as the design is sustainable and elegant too. These usages of 3D printing set an example that it can be used for printing whole cities, rebuilding places which are destroyed due to natural disaster. With this technology, companies can design and develop prototypes and models to judge the uniqueness and layouts of a building saving lot of time and labor of constructor and clients. Some examples like homes are Apies Cor Printed House in Russia [28] and Canal House in Amsterdam.

9.5.6 Textile Industry

There is no sphere of day to day life left which has not been touched by 3D. Today, shoes, jewelry, consumer goods and even fabric [29] are 3D printed. Companies like Adidas and Nike are mass printing g shoes and other sports items which are even customized as per the needs of the customer [30]. The designers and product engineers are using 3D as it is very easy to make innovative designs and print them for trials within a very short interval of time. Also, 3D printing reduces costing and wastages, in turn increasing the revenue of a company.

9.5.7 Electric and Electronics Industry

3D printing in electronics uses MJ technology. A conductive and insulating ink is sprayed on the surface in the form of very thin lines and then hardened using UV rays. There are many benefits of using this technique in electronic industries as it gives inhouse prototyping of circuits and PCBs, faster production and market reach, design flexibility, customization as per requirement and simplified supply chain. Electrodes can be printed by FDM which decreases time and cost and provides flexibility of alteration of design [31]. 3D printing can provide faster rate of production to active electronic components like diodes, rectifiers, operational amplifiers, transistors, light-emitting diodes (LEDs), batteries and so on. These components need complex fabrication methods compared to those required for passive parts having complexity.

9.6 ADVANCEMENT AND FUTURE TRENDS IN 3D PRINTING

3D printing process is an advanced manufacturing technique having wide applications in different areas. The future advancement in this field is mandatory due to high demand in the market. Now this process is used everywhere to manufacture parts from different materials. Researchers are busy in finding material for different applications.

9.7 IS 3D PRINTING READY FOR PRODUCTION?

For many years 3D printing has changed the rules of the game in terms of the rapid prototyping process and shortened the time of product design and development. Rapid development in additive manufacturing has introduced new techniques and materials that are faster, more reliable and provide more satisfying parts. With every improvement the question always arises: is 3D printing ready for mass production? It may never reach full speed and low-cost injection molded parts, but with this technology we are now not limited to specific geometries and materials. Additive manufacturing is taking place today and there are many examples of 3D manufactured parts in various fields. Adidas has sold more than 100,000 of its 3D-printed shoes, which were launched in 2017. The Mini Cooper features a 3D printed dashboard faceplate that users can fully customize.

9.7.1 Bioprinting

Bioprinting is a similar process like 3D printing: it utilizes a CAD file as a base file to print an object step by step. But bioprinting uses biomaterials specific for bio medical and health care applications. Tissue development, organ developments are the examples of this process. Orthopedic implants are also manufactured with this process.

It is very challenging to manufacture components from soft materials, as they hang down without support and lose their structure. A new method known as Suspended Layer Additive Manufacturing (SLAM)uses a hydrogel in which small particles are used to create a self-healing gel. Gel-type fluids or materials can be fed into this environment and built up in layers to form a 3D object.

Recent discoveries show the enormous potential of using 3D printing and scaffolding for tissue engineering. Bioprinting, which involves printing living cells with a specific design, is the most modern way and can produce living organisms. Bioprinting systems are divided into three different methods depending on the technique used (Figures 9.4 and 9.5).

1. The laser technique 2D cell modeling. Laser direct recording (LDW) creates precise patterns of viable cells. These cells are mixed in solution on donor slides and moved to a collector that uses laser energy. The laser pulse forms a bubble, and this generates shock waves. The shock waves help the cells towards the collector in the Petri dishes.

2. Inkjet bioprinting uses live cells that are printed as droplets by the cartridges. This method allows one to print cells which depend upon different parameters.

3. Another way to print cells is extrusion-based printing in which extrusion is done of bio-materials filaments.

Following steps are involved in bioprinting process.

1. **Pre-bioprinting**: In this step a digital file is generated for the use of printer. Today, these files are extracted from MRI and CT scans. After preparation of cells and mixing them with their bio ink, an imaging system is used to ensure the mixing so that printing can be done successfully.

2. **Bioprinting**: Depending upon the bio structure different types of print head are used. As per the requirement different types of inks and bio gels are used to form different tissues and artificial organs.

3. **Post-bioprinting**: Most structures are cross-linked for full stability. Cross-linking is generally accomplished with an ionic solution or

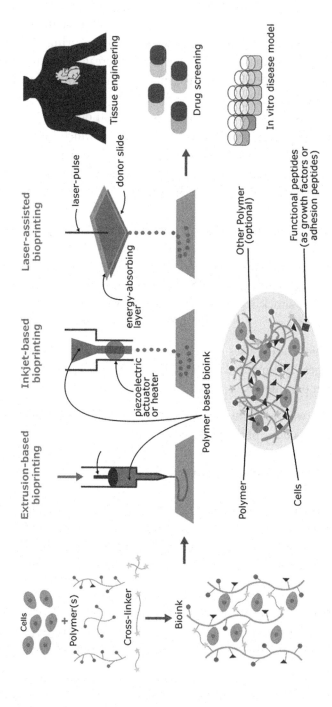

FIGURE 9.4 Different forms of bio printing process.

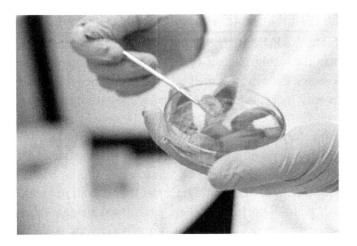

FIGURE 9.5   Mouth tissue part manufactured using bio printing.

UV light—the composition of the construct helps scientists finding what type of cross-link to use. The cell-filled constructs are then put inside the culture incubator.

9.8   APPLICATION OF BIO-PRINTING

Today's bioprinting technologies are still new for different manufacturing sectors. But after some time, it can impact the health care sector with huge benefits.

1. **Drug development**: Much modern research relies on living objects - an inconvenient and costly method for different purposes. It can be utilized in the early stages with a low-cost solution. Researchers can determine a drug candidate's efficacy sooner using bio printed tissues, enabling them to save time and cost also.

2. **Artificial organs**: This is one of the most demanding fields for bio printing. In the case of patients waiting for organ replacements, with this new technique it can be done in a rapid way. Different types of artificial organs can be manufactured by this process.

3. **Wound healing**: Different tissue-related bio inks are available now days, enabling scientists to work with artificial skin, cells, etc. In future in clinics doctors can use these techniques for medical treatments of skin injuries.

9.9 EVOLUTION OF 4D PRINTING FROM 3D PRINTING

3D printing, an additive method, is one of the most groundbreaking innovations in manufacturing. It has changed the scenario of manufacturing along the whole sector. By using this method, we can produce complex shapes and structures with less cost and time. It has witnessed drastic advancements and changes in the field of manufacturing. Despite its ability to manufacture complex shapes it is still not widely acceptable for production.

The growing demand for objects that can be adapted to a variety of applications such as bundling and versatile windmills has contributed to the development of 4D printing. Currently, analysts are seeking to the future with plain 3D printing, where parts from a single fabric create a metamaterial structure. The pattern of metamaterials is formed by the combination of characteristic materials that cause the superimposition of basic reactions when triggered by external shocks. Compatible printing of different materials will create a fabric anisotropy frame, giving rise to the question of changing structure by twisting, lengthening, rotating and folding along the tomahawks. Analysts are helping to work on expansion of these auxiliary changes to develop cabinets, lifts, microtubes, delicate robots, toys, etc. This ability of objects to change their structure over time using distinctive materials is known as 4D printing.

9.10 4D PRINTING MATERIALS AND TECHNOLOGIES

4D print materials are divided upon the external stimuli with which they react. The current group of smart materials are divided into the following categories:

9.10.1 Thermo Responsive Materials

These materials are concerned with the mechanism of the shape memory effect (SME). They are divided into shape memory alloys (SMA), shape memory polymers (SMP), shape memory hybrids (SMH), shape memory ceramics (SMC) and shape memory gels (SMG). Most scientists prefer SMP because printing on these materials becomes easy. They change when heat or thermal energy is used as a stimulus.

9.10.2 Moisture Responsive Materials

It involves materials which are going to interact with water. Such materials are broadly favored by scientists because the availability of water is enough and can be used in different applications. Hydrogel is one of the smart materials which reacts strongly with water. For example, hydrogels can expand by up to 200% of their original volume when they interact with water.

9.10.3 Photo/Electro/Magneto Responsive Materials

It includes materials which interact with light. For example, when photo-sensitizing chromophores are administered in the form of polymer gels at certain locations, they swell and absorb light when they come into contact with natural light. Similarly, when a current is applied to an object containing ethanol, it evaporates, thereby increasing in volume and expanding the entire matrix. Magnetic nanoparticles are embedded in the printed object to obtain magnetic control of the object.

Despite being an advanced technology, 4D printing has to overcome many technological challenges before it can be widely accepted. A few of the challenges in the printing industry include the inability to provide good support structures for complicated objects, no multi-material print-ers, no cheap printers and smart materials, long printing times, and lim-ited reliability for long-run printed objects. While there are some advances in printing technology, such as 5-axis printing devices.

9.11 SUMMARY

3D printing is one of the most popular manufacturing processes in mod-ern days. In this chapter our focus is to explore the new and modern trends in the field of 3D printing. The present work gives insights into the applica-tion of 3D printing in biomedical and food processing fields which is a new and biggest challenge of the future. Materials for artificial organs in 3D printing techniques is a new area of interest which needs to be explored. Based on 3D printing techniques the development of 4D printing process is undergoing which is the scope for future research in this field.

REFERENCES

1. A. Ambrosi & M. Pumera, "3D-printing technologies for electrochemical applications, *Chemical Society Reviews*, Vol. 45, No. 10, pp. 2740–2755, 2016.
2. A. M. T. Syed, P. K. Elias, B. Amit, B. Susmita, O. Lisa, & C. Charitidis, "Additive manufacturing: Scientific and technological challenges, market uptake and opportunities," *Materials Today*, Vol. 1, pp. 1–16, 2017.
3. C. Silbernagel, "Additive Manufacturing 101–4: What is material jetting?" *Canada Makers*, 2018. [Online]. Available: http://canadamakes.ca/what-is-material-jetting/. [Accessed 2019].

4. J. Balli, S. Kumpaty, V. Anewenter, ASME. Continuous liquid interface production of 3D Objects: An unconventional technology and its challenges & opportunities, in: *ASME 2017 International Mechanical Engineering Congress and Exposition, American Society of Mechanical Engineers*, 2017. V005T06A038-V005T06A038.

5. G. Liu, Y. Zhao, G. Wu, & J. Lu, "Origami and 4D printing of elastomer-derived ceramic structures," *Science Advances*, Vol. 4, No. 8, p. eaat0641, 2018.

6. L. Ze-Xian, T.C. Yen, M. R. Ray, D. Mattia, I.S. Metcalfe, & D. A. Patterson, "Perspective on 3D printing of separation membranes and comparison to related unconventional fabrication techniques," *Journal of Membrane Science*, Vol. 523, No. 1, pp. 596–613, 2016.

7. M. A. Caminero, J. M. Chacon, I. Garcia-Moreno, & G. P. Rodriguez, "Impact damage resistance of 3D printed continues fibre reinforced thermoplastic composites using fused deposition modelling," *Composite Part B: Engineering*, Vol. 148, pp. 93–103, 2018.

8. W. Xin, J. Man, Z. Zuowan, G. Jihua, & H. David, "3D printing of polymer matrix composites: A review and prospective," *Composites Part B*, Vol. 110, pp. 442–458, 2017.

9. J. H. Martin, B. D. Yahata, J. M. Hundley, J. A. Mayer, T. A. Schaedler, & T. M. Pollock, "3D Printing of high-strength aluminium alloys," *Nature*, Vol. 549, No. 7672, pp. 356–369, 2017.

10. L. Hitzler, F. Alifui-Segbaya, P. William, B. Heine, M. Heitzmann, W. Hall, M. Merkel, & A. Ochner, "Additive manufacturing of cobalt based dental alloys: Analysis of microstructure and physicomechanical properties," *Advances in Materials Science and Engineering*, Vol. 8, pp. 1–12, 2018.

11. T. DebRoy, H. L. Wei, J. S. Zuback, T. Mukherjee, J. W. Elmer, J. O. Milewski, A. M. Beese, A. Wilson-Heid, A. De, & W. Zhang, "Additive manufacturing of metallic components-Process, structure and properties," *Progress in Materials Science*, Vol. 92, pp. 112–224, 2018.

12. F. Trevisan, F. Calignano, A. Aversa, G. Marchese, M. Lombardi, S. Biamino, D. Ugues, & D. Manfredi, "Additive manufacturing of titanium alloys in the biomedical field: Processes, properties and applications," *Journals Indexing & Metrics*, Vol. 16, No. 2, pp. 57–67, 2018.

13. D.J. Horst, C.A. Duvoisin, & R.A. Viera, "Additive manufacturing at Industry 4.0: A review," *International Journal of Engineering and Technical Research*, Vol. 8, No. 8, pp. 1–8, 2018.

14. W. Haoa, Y. Liua, H. Zhouc, H. Chenb, & D. Fangb, "Preparation and characterization of 3D printed continuous carbon fiber reinforced thermosetting composite.", *Polymer Testing*, Vol. 65, pp. 29–34, 2018.

15. F. Baldassarre, & F. Ricciardi, "The additive manufacturing in the Industry 4.0 era: The case of an Italian FabLab," *Journal of Emerging Trends in Marketing and Management*, Vol. 1, No. 1, pp. 1–11, 2017.

16. D. Owen, J. Hickey, A. Cusson, O. I. Ayeni, J. Rhoades, D. Yifan, W. Limmin, P. Hye-Yeong, H. Nishant, P. P. Raikar, Yeon-Gil, & Z. Jing, "3D printing of ceramic components using a customized 3D ceramic printer," *Progress in Additive Manufacturing*, Vol. 1, pp. 1–7, 2018

17. A. Zocca, P. Lima, & J. Günster, "LSD-based 3D printing of alumina ceramics," *Journal of Ceramic Science and Technology*, Vol. 8, No. 1, pp. 141–148, 2017.

18. T. Lanko, S. Panov, O. Sushchyns'ky, M. Pylypenko, & O. Dmytrenko, "Zirconium alloy powders for manufacture of 3D printed particles used in nuclear power industry," *Problems of Atomic Science and Technology*, Vol. 1, No. 113, pp. 148–153, 2018.

19. T. Xueyuan, & Y. Yuxi, "Electrospinning preparation and characterization of alumina nanofibers with high aspect ratio," *Ceramics International*, Vol. 41, No. 8, pp. 9232–9238, 2017.

20. L. Jian-Yuan, A. Jia, & K. C. Chee, "Fundamentals and applications of 3D printing for novel materials," *Applied Materials Today*, Vol. 7, pp. 120–133, 2017.

21. P. Singh, & A. Raghav, "3D food printing: A revolution in food technology," *Acta Scientific Nutritional Health*, Vol. 2, No. 2, pp. 1–2, 2018.

22. V. Sreehitha, "Impact of 3D printing in automotive industry," *International Journal of Mechanical and Production Engineering*, Vol. 5, No. 2, pp. 91–94, 2017.

23. M. Petch, "Audi gives update on use of SLM metal 3D printing for the automotive industry," *3D Printing Industry*, 2018. [Online]. Available: https://3dprintingindustry.com/news/audi-gives-update-use-slm-metal-3d-printing-automotive-industry-129376/. [Accessed 2019].

24. Z. Liu, M. Zhang, B. Bhandari, & Y. Wang, "3D printing: Printing precision and application in food sector," *Trends in Food Science & Technology*, Vol. 2, No. 1, pp. 1–36, 2017.

25. L. Lili, M. Yuanyuan, C. Ke, & Z. Yang, "3D printing complex egg white protein objects: Properties and Optimization," *Food and Bioprocess Technology*, Vol. 1, pp. 1–11, 2018.

26. J. Norman, R.D. Madurawe, C.M. V. Moore, M.A. Khan, & A. Khairuzzaman, "A new chapter in pharmaceutical manufacturing: 3D-printed drug products," *Advance Drug Delivery Review*, Vol. 108, pp. 39–50, 2018.

27. A. D. Mori, M. P. Fernández, G. Blunn, G. Tozzi, & M. Roldo, "3D printing and electrospinning of composite hydrogels for cartilage and bone tissue engineering," *Polymers*, Vol. 10, No. 285, pp. 1–26, 2018.

28. M. Sakin, & Y. C. Kiroglu, "3D printing of buildings: Construction of the sustainable houses of the future by BIM," *Energy Procedia*, Vol. 134, pp. 702–711, 2017.

29. L. Gaget, "3D printed clothes: Top 7 of the best projects," *Sculpteo*. 2018. [Online]. Available: https://www.sculpteo.com/blog/2018/05/23/3d-printed-clothes-top-7-of-the-best-projects/. [Accessed 2019].

30. S. Horaczek, "Nike hacked a 3D printer to make its new shoe for elite marathon runners," *Popular Sciences*, 2018. [Online]. Available: https://www.popsci.com/nike-3d-printed-sneakers. [Accessed 2019].
31. Y. F. Chuan, N. L. Hong, M. A. Mahdi, M. H. Wahid, & M. H. Nay, "Three-dimensional printed electrode and its novel applications in electronic devices," *Scientific Report*. Vol. 1, pp. 1–11, 2018.

CHAPTER 10

Multimedia Big Data

The Multimedia Big Data Impact on Data Storage, Management and Security Strategies for Data Analytics

Fatima Ziya, Meenu Shukla,
R. Sharmila, and Umang Kant
Krishna Engineering College

Jawwad Zaidi

Dr. Akhilesh Das Gupta Institute of Technology & Management

CONTENTS

DOI: 10.1201/9781003121466-10

10.1 INTRODUCTION

In the course of recent years, huge quantum of information has spread worldwide with the exorbitant utilization of different advanced administrations that constantly create various measures of heterogeneous, organized or unstructured information which is ordinarily known as Big Data [1]. Big Data refers to quick and broad utilization of multimedia data such as text, audios, videos and images and this has been increasing steadily [2]. According to the current situation, information has been increasing exponentially since 2011 [3]. It is anticipated that information will augment multiple times from 130 exabytes in 2005 to 40,000 exabytes in 2020 [3]. Frequent sources that use MMBD are Facebook, YouTube, Twitter, Instagram and many more. For instance, each moment, individuals are transferring approximately 10 h recordings in YouTube every day, clients send roughly 500 million messages in Twitter, and almost 20 billion photographs are transferred on Instagram consistently [2]. As per this investigation, every individual in this world transfers almost 5,200 GB of information.

Because of this innovative development and breakdown of this perplexing information, it is trying to safely store, oversee and share complex enormous measure of information. Enormous information administrations handle such sort of troubles, and the administrations are chief productivity, key choices, income, great administrations, characterizing needs, recognizing new innovations and creating new items, all shrouded in information science [2]. Therefore, to deal with the huge measure of complex information, the multimedia large information gives more broad and progressed arrangements. In addition, this multifaceted nature of information spurs the improvement of the cutting edge executive's

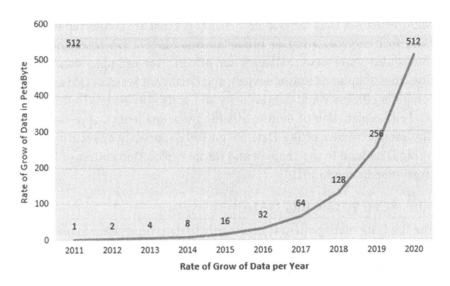

FIGURE 10.1　The prediction of data from 2011 to 2020.

advancements just as methods for managing the difficulties of large information [1]. Figure 10.1 shows the prediction of Big Data information from 2011 to 2020.

Big Data is an assortment of complex information which is difficult to manage, store, visualize, share and protect with current innovations [3]. There are five fundamental attributes to deal with complex enormous information: these incorporate voluminous information from various sources, nature of information (organized or unstructured information), fast pace of given information, changeability of information and accuracy of information. Enormous data characteristics are valuable in isolating the covered plans. Big Data is moreover divided into ten sub-arrangements to the extent data type, data source, data plan, data purchaser, data usage, data assessment, data store, data repeat, data planning, organizations and data dealing [4]. Multimedia services and applications require many possibilities to compute unstructured Big Data. In 2020, it is anticipated that (4*10^24 bytes) information might be created. According to the latest research led by CISCO and International Business Machines, 2.5 quintillion bytes of information are delivered consistently which is almost equivalent to 5,200 Gigabyte for each individual on the planet. Most of the data is produced by the Internet of things (IoT) devices such as sensors, transducers, home appliances and most widely used social media.

Multimedia Big Data develop new challenges that are connected with Big Data such as integrating or miscellaneous MMBD [5]. Internet-related things also share new challenges for MMBD, for example, data collection from ubiquitous sensor devices, and Quality of Service (QoS), assures computer efficiency and data security and scalability issues. In this chapter, brief explanation of multimedia Big Data, challenges, applications of Big Data, attributes of Big Data 5V and 10V's, privacy or security issues in Big Data tend to the chances and future exploration course of MMBD large information in IoT.

10.2 REVIEW ON BIG DATA

The Big Data management is necessarily to determine, data management techniques, characteristics of huge information, challenges, devices and stage including capacity pre-handling security and protection issues. Big Data management also enrolls information warehousing, information quality, data incorporation and information administration [6]. Big Data is a term for composite data especially, where infinite data are arising from multiple origin, are commonly used for digital marketing, business intelligence and decision making [4]. The main goal of huge data management [7] is to build the quality of data, accessibility for decision making as well as well as improve productivity. Due to the advancement in technologies, huge information has gotten a key to the achievement of numerous ventures, science, designing and improvement field. As indicated by the most recent overview, the huge measure of information has expanded in various fields in the course of many recent years. In each long term, the measure of information has been twice since 2011 [4]. In 2003, ($5*10^{60}$ bytes (Exabyte)) of information were produced by human and today this measure of data is made in 2 days. In 2012, digital measure of information in the overall was extended to $2.72*10^{21}$ zettabytes. According to International Data Corporation (IDC) diagram, report and thought about that the colossal amount of data created and grew rapidly within 5 years worldwide to $01.8*(10^{21}$ bytes) zettabytes (ZB) of data [4]. With the short duration, this voluminous information has been multiplying somewhere around each long term. A great deal of huge data started from IoT contraptions. The proportion of data delivered by IoT contraptions is reached out to show up at 600 zettabytes for every year by 2020 [3]. Figure 10.2 depicts the process flow of Big Data management.

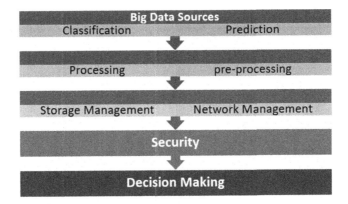

FIGURE 10.2   Process flow of Big Data management.

Multimedia data play an important role on web spine traffic and is increasing by 70% by 2013 [8]. For example, Facebook produces over 10*1015 Bytes of information consistently. Only Google has got more than 1,000,000 servers around the world and it deals with 100 s Petabyte for online searching [9]. For a large amount of data processing and manage hardware community, Google has introduced Map Reduce framework [3]. Apache's Hadoop dispersed document framework is a software segment for distributed computing alongside coordinated part, for example, Map Reduce. Hadoop is an open-source foundation of Google Map Reduce, including dispersed record framework. It gives the component of reflection of the guide and decrease [3]. Moreover, Alibaba produces 10^30 bytes of information every day [10]. With the advancement of the new technologies like IoT, it constantly and significantly generates huge amount of data frequently. For instance, in YouTube, individuals are transferring a normal of 72 h recordings every moment. At present, Big Data management comes with the new techniques in terms of data integrity, data complexity and data storage and data trustworthiness.

10.3 BIG DATA MULTIMEDIA LIFE CYCLE

Information generation is the primary phase of the multimedia life cycle model. The most widely used example of multimedia Big Data is web data. Due to the excessive use of internet, it is originated from chat records, uploading videos on YouTube, uploading pictures on Facebook, Instagram, blog messages, and forum posts [11]. These data are growing steadily because of individuals' lives, which is created from particular sources

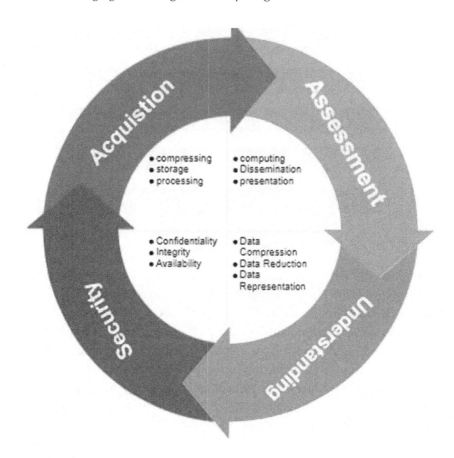

FIGURE 10.3   Life cycle of Big Data.

like camera click, sensors, videos etc. The difficulties of multimedia Big Data are detecting information, data generated from scientific research and so on. Most of the multimedia data are generally created from IoT devices, which is the essential goal of Big Data. It has many applications produced through IoT-enabled smart cities and agriculture, industries, traffic monitoring, healthcare field etc. [4]. Figure 10.3 depicts the lifecycle of Multimedia Big Data Management.

10.3.1 Data Acquisition

Acquisition of information is the main period of Multimedia lifecycle model. It gets interactive media information from the heterogeneous sources such as web, IoT devices, sensors, videos, digital games, audio and so forth. In latest research, users act as a monitor and transfer information at runtime [12]. Recently, numerous guidelines for video coding have been explored. It has been extremely testing to getting the information from numerous sources because of unstructured method of information portrayal. The ambiguous amount of data builds lots of volume, size and quality. These features of interactive media can arrange new methodologies and techniques to manage intricate and heterogeneous information. Table 10.1 shows the contrast between multimedia data set with ordinary data sets.

- **IoT multi-media Big Data generation**: In Big Data multimedia, various IoT applications have an extensive use, for instance, healthcare, satellite, smart cities/nations and education. With the advancement of IoT technologies, it is beneficial for MMBD [13] in terms of optimizing, computing data and storage. To produce multimedia information from IoT gadgets or transferring information, the association layer is isolated into distinct layers, for instance, the physical (identifying) layer, application layer, and the organization layer [4]. The obtained information is executed by the detecting layer, which comprises the organization layer. Transmission of data and preparing are done by the organization layer. The sensor network is answerable for transferable information in a short period of time, though the transmission of information in a significant distance is conveyed by Internet. The application administration of the IoT is conveyed with the assistance of the use layer. Table 10.1 shows the different

TABLE 10.1 MULTIMEDIA DATA SETS WITH ORDINARY DATA SETS

Characteristics	Data Sets	Big Data	Multimedia Big Data
Volume	Low	Medium	High
Data size	Definite	Uncertain	Big
Inferring video	Not Possible	No	Yes
Representation of data	Structured data	Structured data	Unstructured data
Real time	Not possible	Yes	Yes
Human-centric	Not possible	No	Yes
Data source	Centralized	Heterogeneous distributed	Heterogeneous distributed
Complexity	Low	Medium	High
Response	Not possible	No	Yes

quality of mixed media huge data. IoT applications are generally new for MMBD, which is halfway charge by expanding scope of detecting gadgets and availability of MMBD [14].

10.3.2 Data Compression/Reduction

With the exponential growth of MMBD, various data compression techniques are used for efficient storage, less consumption of memory and communication [15]. In data reduction or compression of data, irrelevant information is eliminated, and it refers to dispose the excess information from data sets. Redundant data means duplication of data, use common sub-expression elimination method, redundancy affect system inconsistency, storage, transmission cost, and data reliability. Various data compression techniques are as follows:

- **Feature transformation-based data compression:** The target of this feature transformation is to minimize mathematical information utilizing basic sign preparation or transform technique [14]. Examples of feature transmission data reduction are compressing sensing and wavelet transform.

- **Analysis-aware compression:** The objective of this compression technique is the decrease of excess, inconclusive, indefinite or insignificant picture while protecting the perceptual quality of the image. Examples of these pressure techniques include (JPEG) Joint Photographic Experts group or lossy image compression algorithm that uses unwanted compression to the reconstructed image [12].

- **Cloud-based compression:** Because of the environment of IoT, immense measure of MMBD is created. In huge associations, the voluminous measure of data is stored in cloud, which prompts stockpiling issues in distributed computing. Some stockpiling-related issues are access control, space time, approval and so forth. In Facebook, in any event 300 million pictures are transferred every day, though research by Microsoft has looked at it, distributed storage administration obliges around 11 billion pictures. On the cloud side, cloud-based huge information pressure procedures are accessible with respect to reality to store media information. In a PC society, distributed computing stockpiling administrations are continuously utilized for sight and sound enormous information [12].

10.3.3 Data Representation

The data received by multimedia data comes from distinct sources and the representation of each source of data is different. It tends to have different representations for each source, or sometimes common representation occurs. This exponential data representation is commonly described by both the descriptive and structural metadata [16]. Multimedia representation of data comprises the following methods.

- **Feature-based data representation**: The aim of feature-based representation is to search the best representation including all feasible combinations. Some features of MMBD are standard regarding time and space [17]. Feature-based representation impose the information among all possible combination of attributes. With the assistance of this application, audio, video, stream or picture pixel, all highlights are separated and consolidated into vectors. This technique drives the adaptability and precision need highlight-based representation.

- **Learning-based representation**: In this portrayal, the information is removing the shrouded space which is called learning or machine-based portrayal [18]. There are various deep learning methods to perform multimedia Big Data representation, such as deep Auto encoder, Deep Boltzmann machines such novel algorithms aims to learn high-level representation from low-level features using a series of non-linear transformations [14]. Today, deep architecture is widely used for learning-based representation.

10.3.4 Processing and Analysis of Data

After the information is gained or put away, the following period of the existence cycle is information preparing and investigation. Interactive media crude information, which is obtained from the various stages are unorganized, uncertain or loud information. The ambiguous sizeable interactive media data sets are not straightforwardly appropriate for examining or processing because the acquired data is noisy, sparse, distinct data which causes troublesome and unfeasible problem [19]. The objective of data processing is to convert raw or useless data into new and clean data. After processing the data, the data sets are prepared for additional high level analysis. The steps of multimedia data preparing are data cleaning, data transformation and data reduction [12].

- **Data cleaning**: As indicated by the specialists, information researchers are investing their vast majority of time in information sorting out and cleaning. The main steps used for organizing and cleaning the data are noise reduction, acquisition, eliminate irrelevant information, maintain consistency of data and outlier identification. Data cleaning can enhance the quality of data as well as discrepancy and faultiness of data. With the assistance of information control and data processing tools, the noisy semi-organized data is transformed into clean data [16].

- **Data transformation and integration**: The objective of data integration is to combine the relevant stages as well as their metadata into a steady source. It is the process to detect the faultless and resolve it. However, transformation of data is the main step in data processing. Transformation of data includes data formatting, aggregation and decomposition. In the latest research, various new algorithms are applied to build a similar representation of model to transform distinct data [20].

- **Data mining**: Numerous scientists have been inspired by the well-known Deep Learning tool stash to remove huge scope highlights utilizing Deep learning algorithms. Deep learning has predominantly summarized the large volume of data into modules, where the common features are placed together. The objective of clustering-based algorithms is to gather large volume of data and compress it into nearly small and minimum storage resources. As the capacity of data grows along with variability size and volume, it is impossible to store the high volume of future data. Due to the advancement of Big Data, many researchers design a novel algorithm for maintaining the storage structure. Buza Nagy et al. 2014 [21] proposed Storage-Optimizing Hierarchical Agglomerative Clustering (SOHAC), a calculation for planning a capacity structure that offers a diminished extra room necessity for variable data [22].

- **Artificial Intelligence (AI)**: AI is the innovation of enhancing the exhibition of personal computer programs by learning the information naturally. AI can arrange into supervised and unaided learning. The supervised learning uses an approach to learn the mapping function from the given input to output [4], whereas in unsupervised learning, there are no approaches and no label related to each data [1].

- **Data depletion**: Different data compression techniques are used for information depletion. There are two types of reduction techniques: instance depletion and feature depletion. In immediate reduction, the quality of data is improved by reducing the conflict and minimizing the complexity; instance depletion maintains the originality of data and integrity of data. Feature extraction conquers the hole between low-level interactive media attributes into significant level qualities. It extracts the features from voluminous data set, the whole is time consuming [14].

10.3.5 Storage and Retrieval of Multimedia Data

To store the large amount of information, MMBD require data base management system which is known as Multi-Media Database Management System (MMDBMS) [23]. It consists of multimedia information as well as relationship. The attributes of MMDBMS are capacity, requirements on spatial and transient, introduction of information, recover and so on.

- **Modeling of multimedia data**: There are various database modeling such as relational database, network modeling, semantic, but multimedia supports only few due to the unstructured nature MMBD. The demonstrating framework for the MMBD archive, which joins the advancements, for example, Database Management for Object Oriented System, Natural Language Processing (NLP), and so on, to select the crucial data, structure the information reports and offers syntactic recovery. The information demonstrating is basically used to separate/recover the data.

- **High volume storage management**: The capacity of interactive media depicted by critical volume and assortment which need a various leveled structure. The leveled structure of MMBD increases huge information, which builds the capacity size and diminishes the presentation.

- **Performance**: Performance is an important feature of MMBD, in terms of consistency, accuracy, completeness, generating of data on continuous time and execution. QoS and Quality of Experience (QoE) are the examples of MMBD presentation.

- **Multimedia indexing**: Due to unstructured information design, the traditional RDBMS isn't suitable for Big Data multimedia. This issue is resolved with the assistance of indexing. Indexing strategies have been proposed to deal with the diverse information types and inquiries as well as to easily extract the data.

10.3.6 Data Assessment

Advancement of data innovations and MEMS (Micro Electro Mechanical Sensor) innovations and its broad development in various regions brought about a colossal measure of various information, for example, recordings, sounds, and text information. Because of the fast improvement of Big Data multimedia information and administration, it is imperative to provide the (QoE) to the clients.

10.3.7 Data Computing

From the tremendous measure of mixed media information, it is a tough and challenging task to sort the large amount of data in an organized manner. Big Data multimedia figuring is a novel technique; the information investigation is performed by joining huge scope calculation with numerical models.

10.4 CHARACTERISTICS OF MULTIMEDIA BIG DATA

The MMBD has numerous difficulties when contrasted with the customary content-based huge information regarding fundamental tasks, for example, enormous volume of data sets, handling, transmission and investigation of information. Figure 10.4 shows the characteristics of Big Data Multimedia Management or 5V's of Big Data. A portion of the difficulties of multimedia Big Data is given as follows:

- **Real-time data access and its quality**: For the most part, the MMBD is prepared and examined continuously. To accomplish the QoE, the information should be progressively, investigated the information in equal/appropriated way for examination, learning and mining.

- **Unorganized and heterogeneous data**: The presentation and modeling of Big Data multimedia are difficult due to indefinite, heterogeneous and multimedia data. For example, the conversion of data from unstructured data to structured data. How to represent or model the data from different sources?

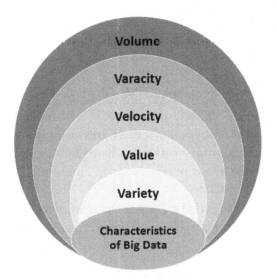

FIGURE 10.4   Different characteristics of Big Data multimedia management.

- **Complexity of understanding and cognition**: Due to the sematic gap, it is difficult for the computer to understand the sematic gap between the high level and low level features of Big Data multimedia. MMBD are progressing with respect to time and space.

- **Scalability and competence of Data**: Big Data Multimedia consists of large volumes of different data types which needs large-scale computation. It needs more computation, storage and communication resources.

The aforementioned challenges lead to the different complications that are mentioned as follows:

1. **Computation of data**: How efficiently the data mining and learning is performed to examine the data?

2. **Optimization of computing, storage and communication**: How multimedia architecture is designed to use the storage, processing and communication efficiently?

3. **Real-time computation**: How to perform online processing on entire multimedia Big Data in parallel or distributed manner to satisfying the QoE?

4. **Presentation and modeling**: How would we set up the portrayal and demonstrating the unstructured, multimodal and heterogeneous multimedia data?

10.5 APPLICATIONS IN MULTIMEDIA BIG DATA

The organization of Big Data mixed media framework depends on the commonplace large information procedures to examine, measure and control the multimedia information productively. A portion of the utilization of mixed media procedures is shown in Figure 10.5.

- **Internet of Things (IoT)**: The Big Data Multimedia has a wide range of IoT applications such as healthcare, smart cities and satellite imaging.

- **Medical or healthcare applications**: The MMBD has transformed the field of medical, healthcare, business, society and engineering science. Due to the development of MEMS technology, the size of wearable sensor devices is reduced which offers uninterrupted physiological examination. The data collected form the sensors is transmitted to a gadget and stored in cloud by the service provider. Kumari et al. [6] studied fog computing, cloud computing and IoT

FIGURE 10.5 The application of MMBD.

for healthcare services. It is extremely fundamental to comprehend the Multimedia Big Data from the IoT point of view to understand the validation of information to provide appropriate diagnosis.

- **Social networks**: A great deal of examination works/investigation has been done on the interpersonal organization. Many researchers have examined the difficulties of social exercises and practices of individuals on Twitter hash label which has an enormous number of data sets, and simple entry. The social recommender framework is an arising innovation which is basically used to share media data.

- **Surveillance videos**: One of the assuring sources of Big Data multimedia is surveillance video [3]. Proposed the idea of recognizing semantic concepts in surveillance recordings [6], as well as many novel Big Data multimedia solutions derived from surveillance sources such as car surveillance, criminal investigation, and IoT.

- **Other applications**: The utilizations of Big Data multimedia can be characterized as biomedicine data, healthcare data, and disaster management system. The biomedical information is the primary origin for MMBD which has a distinct variety of data such as patients medical records, medical images, and patient prescription.

10.6 BIG DATA SECURITY AND PRIVACY APPROACHES

In the age of huge information, individuals exploit internet providers. To examine the voluminous measure of information will be more mind boggling and hard to oversee; it has an incredible business estimation of web access suppliers. With the quick development of Internet innovations, individuals gather the information from various sources to examine and deal with this enormous measure of information where security has become a genuine concern [24]. According to Gartner, 80% of affiliations will confront difficulties by 2016 [1]. The fundamental purpose for this is large information has perplexing information and various qualities. To tie down the huge sum to information, exceptionally proficient techniques and calculations are applied. In this segment primary security segments, including classification, Integrity, Privacy and accessibility will be examined. Figure 10.6 speaks to the ongoing security issues and existing arrangements.

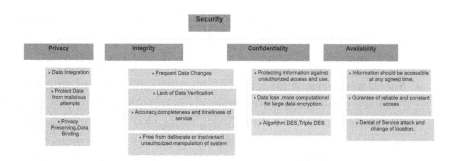

FIGURE 10.6   Data protection requirements.

10.6.1 Data Confidentiality

Confidentiality of data refers to the data protection from unauthorized and unintended users. In the context of Big Data the data are stored in databases to maintain the security in database encryption algorithms. Encryption is the process of encoding the information. Encryption can be ordered into various stages such as table encryption, disk encryption and data encryption. Some of the difficulties related to Big Data confidentiality are discussed as follows. Due to the rapid development, information and encryption algorithm are simply pertinent limited amount of data, but some novel encryption algorithms are required that are related to Big Data confidentiality. The novel encryption algorithm has minimum cost and complexity [1,24]. Overview of encryption algorithms are shown in Table 10.2.

10.6.2 Data Integrity

Integrity of data refers to protection of data from unauthorized modification and also generalized to data trustworthiness, which refers to making sure that the data are not changed by unauthorized party but also make sure that data are free from errors, data are updated and originating from reputable sources [1]. To maintain the integrity of data is particularly very difficult in large-scale organizations, where data are collected from multiple sources and leaders investigate this information, apply different information mining approaches and settle on basic choices productively [12]. The problem in data integrity is to ensure the trustworthiness of data which often depends on the application domain. Solutions are provided like combining distinct techniques,

TABLE 10.2 Overview of Encryption Algorithms

Algorithm	Developed by	Size of Key	Size of Block	Number of Rounds	Existing Cracks	Structure of Algorithm	Crack
Triple DES [25]	IBM in 1978	112 or 168 bits	64 Bits	48	Theoretically	Feistel Network	No
DES [26]	IBM in 1975	56 bits	16 bits	16	Brute force attack, Linear cryptanalysis	Balanced Feistel Network	Yes
Two Fish [27]	Bruce Schneier 1993	128,192 and 256 Bits	128 Bits	16	Differential cryptanalysis	Feistel Network	No
Blowfish [25]	Bruce Schneier 1993	128 Bits	64 Bits	16	Second-order differential attack	Feistel Network	No
RC4 [28]	Ron Riverst in 1987	40–2,048	Variable	256	Weak Key Schedule	Stream	Yes
RC2 [29]	Ron Riverst in 1987	8–1,024 Bits	64 Bits	16 of type Mixing, 16 of type Mashing	Related Key Attack	Heavy- Feistel Network	Yes

digitally sign the data using cryptographic algorithm, to access the data and modification of data is allowed by only the authorized parties and authenticated users, for automatic a detecting fixing the errors. Without using the integrity, collaboration between the organizations cannot be determine as the handiness of information turns out to be less and the data separated isn't reliable. Many Database Management [21] are allow users to express a wide scope of condition also known as integrity constraints. The main objective of these constraints is to maintain the accuracy and consistency of data. The most significant and broadly used definition of data integrity is preventing illegal or unauthorized modification. In the generation of Big Data the major task related to Big Data is in building up a strategy that can guarantee trustworthiness. In a Big Data, some novel integrity rules and algorithms are designed, and the existing algorithms are no longer applicable owing to new qualities of large information.

10.6.3 Privacy

In the period of Big Data, a major risk is the leaking of information because then information is circulated at an extremely fast pace. An ongoing example of leaked information demonstrates that the scope of gathering wide information, and examined by the National Security Agency (NSA) and other national security agencies, have captured the attention of the public to balance between the privacy and the risk of wide data opportunities [24]. In the field of national security, huge information makes an open door in the field of marketing and analysis. Privacy of data is seen as the same requirement as data confidentiality because if the data are not protected by unauthorized sources then privacy cannot be ensured. However, data privacy have some challenges deriving from the need of taking into account requirements from authenticate privacy regulations, as well as individual privacy preferences. For example, there is a need to keep up a strategy that covers all clients' protection concerns. With the development of the technologies is easy to discover the breaking privacy rules. Some novel techniques are also used to ensure that user data is misused or leaked [26].

With the advancement of society, individuals are more concerned and give increasingly more consideration to keep up information security. The current conservation on security investigates the risks caused by huge scope of information but neglected to reveal the benefit, which are treated

TABLE 10.3   Overview of Privacy Preserving Technique

References No	Proposed Solution	Technique	Descriptions	Solutions
Aggarwal and Aggarwal, 2001 [30]	Reconstruction for Algorithm privacy preserving data mining	Expectation maximization algorithm	Privacy Preserving Measurement	Efficiency of randomization
Squicciarini, Sundareswaran et al. 2010 [31]	Three Tier Data protection	Portable data mining	Use data indexing to address privacy preserving algorithm	Protect data from unauthorized or malicious attempts
Xuyun, Chang et al. 2012[32]	MapReduce framework for privacy preserving	Data privacy preservation	Using MapReduce task to ensure privacy preserving	Integration with other data processing
Xuyun, Chang et al. 2013 [32]	Upper bound privacy leakage	Use Heuristic algorithm for cost reducing privacy preserving	Ensure the integrity of data	Efficiency of data processing

as a combination of private, government and cooperate agency. Numerous western nations have set up extraordinary security assurance to ensure residents protection and data. Security of information [33] additionally helps leaders to comprehend and alleviate the risk with respect to information security. By setting up a security insurance organization the confidentiality is checked, yet additionally inspect the protection on business ventures and governments [34]. Table 10.3 shows the protection saving arrangements, strategies and restrictions.

10.6.4 Availability of Data

With the continuous progression of Big Data era, the voluminous measure of data has expanded essentially. In a cloud where huge measures of information are put away, the accessibility of information is significant for rethinking [2]. Cloud performs various services such as Software services, Platform services and infrastructure as a service; if any of the services is not available to the user, when required, then the QoS does not meet the service level agreement (SLA) [27].

10.6.5 Threats to Data Availability

Explosive development of web insider and outsider threats can cause administration degradation. The outsider threats may cause information inaccessibility, and denial of service is the biggest example of such threats, whereas internal threats can occur because of the malicious user, who want to take the advantage from data transformation and organization. Also, assaults on information accessibility, it is important to create techniques that can guarantee the ID of the aggressor as well as help the casualty to keep up security, when such kind of assault happens. With the help of two common methods such as complete replication and party schemes, the availability of data is achievable. To maintain security on Big Data is a tough task, Security on Big Data is a current research territory. The suitable arrangement of these methodologies is to safeguard the integrity and protection of information.

10.7 CHALLENGES OF MULTIMEDIA BIG DATA

Multimedia Big Data not only reveals all five V's (Volume, Variety, Velocity, and Veracity & Value) of Big Data but also accelerates the difficulties into another level.

1. **Security & Storage of Data**: While multimedia has acquired essentially helpful data, it additionally brings new difficulties, for example, information multi-modalities, adaptability, storage, and security. The examples of enormous measure of recordings required each day by advanced mobile phones, reconnaissance frameworks, police body-worn cameras, firearms furnished with little cameras, etc. Presently, the information gathered by each of these gadgets has an issue of putting away and preparing a lot of information and is computationally sensible and financially savvy. These days, quick headways in information storage and equipment make it conceivable to store and handle a particularly enormous measure of information to produce new experiences and important data. Figure 10.7 addresses the mixed media Big Data and its difficulties.

2. **Ineffective Data Representation**: Another serious issue is to measure the blended MMBD in a proficient way. Another multi-modular data recovery framework is the goal to approve conventional hunt frameworks utilizing data from different sources and modalities.

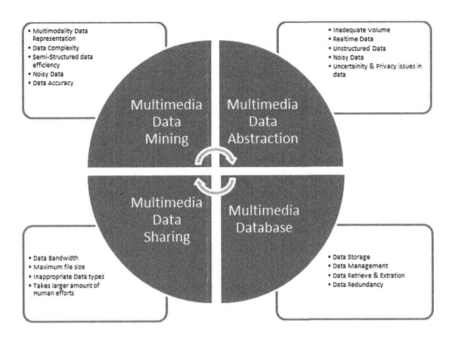

- Multimodality Data Representation
- Data Complexity
- Semi-Structured data efficiency
- Noisy Data
- Data Accuracy

- Inadequate Volume
- Realtime Data
- Unstructured Data
- Noisy Data
- Uncertainty & Privacy issues in data

Multimedia Data Mining

Multimedia Data Abstraction

Multimedia Data Sharing

Multimedia Database

- Data Bandwidth
- Maximum file size
- Inappropriate Data types
- Takes larger amount of Human efforts

- Data Storage
- Data Management
- Data Retrieve & Extration
- Data Redundancy

FIGURE 10.7   Challenges of multimedia Big Data.

3. **Data Knowledge Retrieval**: Another major issue is to quantify the mixed MMBD in a capable manner. Another multi-measured information recuperation structure is the objective to favor customary pursue frameworks utilizing data from different sources and modalities.

4. **Data Security**: The security of information is likewise a significant issue in interactive media huge information examination. New advances, for example, cloud administrations won't hazard or sneak the information protection. Furthermore, wearable gadgets and advanced cells don't protect one's security. This model gives multi-layer access control authorization for various sight and sound segments (e.g., sound or visual fragments) by displaying the spatial-transient requirements and using the MPEG-7 media standard. To have secure access of online MMBD enormous information, another hybrid system is proposed to create and actualize the protection approaches.

5. **Unstructured and Multimodal Data**: In Big Data Multimedia, the major issue is to store the data which is unorganized and multimodal because it is received from miscellaneous sources. Due to this

problem it is always a big gap to reconstruct the unstructured data to structured form.

6. **Inappropriate & Non-Readable Form of Data**: Multimedia information can't be promptly perceived by a PC machine because of significant level and low level semantic hole of data. Therefore, this data may vary from time as well as space.

7. **Big Computational Tasks**: The Big Data Multimedia requires a huge set of calculation and computational task so, there is always a need of high amount of accuracy and efficiency of the data which requires enhancement of computational resources.

10.7.1 Future Discussion

Despite the fact that exploration on enormous information the executives has just accomplished a lot, and there are still a great deal of difficult issues that stay to be unraveled. To assist analysts with improving handle of future examination bearings in the field of enormous information the board, more knowledge into future exploration difficulties and openings are given as follows [2]:

- **Large data pre-processing**: Some significant reasons ensure that information preprocessing difficulties will keep on connecting huge information science for quite a while For example, interpersonal organizations urge individuals to discover, accumulate or produce their own information and offer it with others. They are engaged with urging individuals to create or discover appropriate information, making information sharing simple and offering customers motivating forces [35] to do as such, and furthermore utilizing enticing strategies on information security inclusion and the anticipation of any downsize of information proprietors' framework execution. Information coordination is essentially about getting individuals to team up and share information.

- **Big data analytics**: Huge information investigation identifies with mining, data set looking, and examination devoted to a creative IT capacity to improve organization execution [33]. Two primary objectives of information investigation are to pick up understanding into the connection among highlights and create viable information mining techniques that can precisely foresee future cases. Dissecting huge information straightforwardly suggests enormously tedious route in a tremendous pursuit space to offer rules and criticism to clients.

- **Big data governance**: Enormous information administration is a center part of large information of the board. Information administration characterizes the standards, laws and power over data [28]. For instance, if enormous information is put away in the cloud, it is critical to have a few strategies and decide that characterize which kinds of information should be put away and how rapidly information should be obtained.

10.8 CONCLUSIONS

The requirement for viable administration strategies and innovations to deal with huge information is turning out to be crucial. This study introduces a thorough review of large information of the executives and proposes the administration cycle stream as scientific categorization. Enormous information the board was examined as far as information stockpiling, pre-handling, preparing and security and best in class procedures for every part associated with the administration cycle. Moreover, to sort out how enormous information the board is managing capacity assets, instruments, and innovations for information pre-preparing and handling, we examined the procedures as talked about for large information the executives cycle stream dependent on scalability, availability, trustworthiness, heterogeneity, asset enhancement, and speed [36]. Moreover, we examined the security suggestions on each layer of huge information the board measures. At long last, we presume that enormous information the executives is in its baby stage. Overseeing enormous information requires more proficient methods and instruments to help the pattern of Big Data. Later on, the difficulties and issues tended to in this review will lead the scholarly community and industry to devise better answers for guarantee the drawn out accomplishment of enormous information the executives and to all in all investigate new domains.

REFERENCES

1. Siddiqa, A., I. A. T. Hashem, I. Yaqoob, M. Marjani, S. Shamshirband, A. Gani, F. Nasaruddin. (2016 April). "A survey of big data management: Taxonomy." *Journal of Network and Computer Applications*, 71: 151–156.

2. Kumari, A., S. Tanwar, S. Tyagi, N. Kumar, M. Maasberg, K.-K.R. Choo. (2018 July). "Multimedia big data computing and internet of things applications: A taxonomy and process model." *Journal of Network and Computer Applications*, 124: 169–195.

3. Sharmila, D. K., P. Kumar, and A. Ashok. 2020. Introduction to Multimedia Big Data Computing for IoT, e Library, Volume 163, doi:10.1007/978-981-13-8759-3_1.

4. John, R. et al., Riding the multimedia big data wave. *Proceedings of the 36th International ACM SIGIR Conference on Research and Development in Information Retrieval*, 2013, ACM, Dublin, Ireland, pp. 1–2.

5. Schneier, B., et al. (1999). *The Two Fish Encryption Algorithm: A 128-Bit Block Cipher*, John Wiley & Sons, Inc., USA.

6. Kumari, S., Tanwar, S. Tyagi, N. Kumar. (2018). "Fog computing for healthcare 4.0 environment: Opportunities and challenges." *Computers & Electrical Engineering* 72: 1–13.

7. Thakur, J., N. Kumar (2011). "DES, AES and Blowfish: Symmetric key cryptography algorithms simulation based performance analysis." *International Journal of Emerging Technology and Advanced Engineering* 1(2): 6–12.

8. Inukollu, V. N., S. Arsi, S. R. Ravuri. (May 2014). "Security issues associated with big data in cloud computing." *International Journal of Network Security & Its Applications (IJNSA)* 6(3): 45.

9. Barnes, T. J. (2013). "Big data, little history." *Dialogues in Human Geography* 3(3):297–302.

10. Terzi, D. S., R. Terzi, S. Sagiroglu. A survey on security and privacy issues in big data. *The 10th International Conference for Internet Technology and Secured Transactions (ICITST-2015)*, 2015.

11. Sudeep, T., Sudhanshu, T., Neeraj, K. (Editors). Multimedia Big Data Computing for IoT Applications, Volume 163 ISBN 978-981-13-8759-3 (eBook). doi:10.1007/978-981-13-8759-3, India.

12. Sena, D., Ozturkb, M., Vayvayc, O. (2016). "An overview of big data for growth in SMEs." *Procedia-Social and Behavioral Sciences*, Turkey 235: 159–67.

13. Yun-Peng, Z., et al. Digital image encryption algorithm based on chaos and improved DES. *IEEE International Conference on, Systems, Man and Cybernetics*, India, 2009. SMC 2009.

14. Lv, Z., X. Li, H. Lv, W. Xiu. (2019). "BIM big data storage in WebVRGIS." *IEEE Transactions on Industrial Informatics* 16(4): 2566–2573.

15. Zhao, W., et al. Parallel k-means clustering based on mapreduce. *IEEE international conference on Cloud Computing*, 2009, Springer, Berlin, Heidelberg, pp. 674–679.

16. Zhu, W., P. Cui, Z. Wang, G. Hua. (2015). "Multimedia big data computing", *IEEE Multimedia*, 22(3): 96–c3.

17. Zhang, M. Strict integrity policy of Biba model with dynamic characteristics and its correctness. *International Conference on Computational Intelligence and Security*, China, 2009. CIS'09.

18. Michael, K., K. W. Miller (2013). "Big data: New opportunities and new challenges." *Editorial: IEEE Computer* 46(6): 22–24.
19. Agrawal, D., et al. "Data management challenges in cloud computing infrastructures." *International Workshop on Databases in Networked Information Systems*, 2010, Springer, Berlin, Heidelberg: pp. 1–10.
20. Squicciarini, A., et al. Preventing Information Leakage from Indexing in the Cloud. *IEEE 3rd International Conference on Cloud Computing (CLOUD)*, Miami, FL, USA, 2010.
21. Fluhrer, S., et al. Weaknesses in the key scheduling algorithm of RC4. *International Workshop on Selected Areas in Cryptography*, 2020, Springer, Berlin, Heidelberg.
22. Benjelloun, F. Z., A. A. Lahcen, S. Belfkih. An overview of big data opportunities, applications and tools. *Intelligent Systems and Computer Vision (ISCV)*, 2015.
23. Zhang, X., C. Liu, S. Nepal, S. Pandey, J. Chen (2013). "A privacy leakage upper bound constraint-based approach for cost-effective privacy preserving of intermediate data sets in cloud." *IEEE Transactions on Parallel and Distributed Systems* 24(6): 1192–1202. doi: 10.1109/TPDS.2012.238.
24. Selimović, A., B. Meden, P. Peer, A. Hladnik. Analysis of content-aware image compression with VGG16. *IEEE International Work Conference on Bioinspired Intelligence (IWOBI)*, San Carlos, Costa Rica, 2018.
25. Buza, K., et al. (2011). A distributed genetic algorithm for graph-based clustering. *Man-Machine Interactions 2*, Springer, Germany: pp. 323–331.
26. Sebé, F., et al. (2008). "Efficient remote data possession checking in critical information infrastructures." *IEEE Transactions on Knowledge and Data Engineering* 20(8): 1034–1038.
27. Hu, J., A. V. Vasilakos. (September 2016). "Energy big data analytics and security: Challenges and opportunities." *IEEE Transactions on Smart Grid* 7(5): 2423–2436.
28. Chen, D., H. Zhao. Data security and privacy protection issues in cloud computing. doi: 10.1109/ICCSEE.2012.193.
29. Wang, M., et al. (2013). "High volumes of event stream indexing and efficient multi-keyword searching for cloud monitoring." *Future Generation Computer Systems* 29(8): 1943–1962.
30. Zhao, Z., W. Ng. A model-based approach for RFID data stream cleansing. *Proceedings of the 21st ACM International Conference on Information and Knowledge Management*, China, 2012, ACM.
31. Zografos, K. G., K. N. Androutsopoulos (2004). "A heuristic algorithm for solving hazardous materials distribution problems." *European Journal of Operational Research* 152(2): 507–519.
32. Agrawal, D., C. C. Aggarwal. On the design and quantification of privacy preserving data mining algorithms. *Proceedings of the Twentieth ACM SIGMOD-SIGACT-SIGART Symposium on Principles of Database Systems*, 2001, Santa Barbara, California, USA, pp. 247–255.

33. Bertino, E., Ferrar, E. (2018). *Big Data Security and Privacy*, Springer International Publishing AG.
34. Zhang, D. Big data security and privacy protection. *8th International Conference on Management and Computer Science (ICMCS 2018)*, China 2018.
35. Seref, S., S. Duygu. Big data: A review. *International Conference on Collaboration Technologies and Systems (CTS)*, 2013.
36. Ibrahim, A., et al. (2015). "The rise of "big data" on cloud computing: Review and open research issues." *Information Systems* 47: 98–115.

CHAPTER 11

Voice Assistants and Chatbots Hands on Essentials of UI and Feature Design Development and Testing

Pavi Saraswat, Bharat Bhardwaj, and Prashant Naresh
AKTU

Alaknanda Ashok
GBPUAT

Raja Kumar
Taylor's University

Manish Kumar
CGC Landran

DOI: 10.1201/9781003121466-11

217

CONTENTS

11.1 INTRODUCTION OF INTELLIGENT VIRTUAL ASSISTANTS: CHATBOTS AND VOICE ASSISTANTS

An intelligent virtual assistant or intelligent personal assistant is a software agent who usually performs specific tasks or provides services for a command given to it or on individual basis. So basically, it is an application that comprehends natural human language and completes the electronic task for the specified user either through text or voice. And the basic aim of the intelligent agent is to present all the information that can be retrieved from the internet to complete the task. The intelligent agent gets completed tasks on a daily basis to improve their skills and find the best solutions for the problems.

11.1.1 Chatbots Introduction

As technology grows, the constant need for the intelligent agent has arisen. The main aim of chatbots was to provide service in the form of serving the queries from the clients. Today, chatbots have made a big hype over the world wide web as the organizations can integrate this technology into their businesses and make the most of it. They are widely applicable in all sorts of businesses and industries as it gives the best connectivity between the organization and the clients.

The basic building block of the chatbot is natural language processing (NLP). NLP is the technology that takes the voice message and recognizes it just like Google's Now, Apples' Siri and Microsoft's Cortana. Figure 11.1 explains the process of NLP. Chatbot basically handles the information given to it may it be any form (textual or voice) from the user before recognizing the series of complex algorithms which a message delivered by the user contains [1]. That message contains the information what the user wants to say and how the user will react or fulfill his needs.

A new breed of software with a great potential for a wide range of web applications is being represented by the intelligent agents [2]. Initially, they served as the intelligent user interfaces, personal assistants and an intelligent way to manage mails, client queries and more. Later, the chatbot

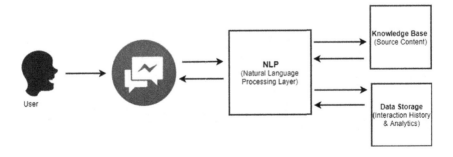

FIGURE 11.1   Natural language processing process.

brought the revolution in the E-Commerce industry, and it was implemented on each and every e-commerce platforms, and the performance was improved abruptly. It affected both business aspects Business-to-Commerce (B2C) as well as Customer-to-Customer (C2C).

It was a revolution because their task was to answer the client's inquiries whole day and night even after the working hours of the employees and there was no limit to questions and the response time was very less. There is a wide range of applications of chatbots which includes call center exchange, banking, marketing, healthcare, language translation, traveling business, real estate industry, fashion industry and many more [3]. And it can be integrated with multiple platforms such as Facebook messenger, Viber, WhatsApp, WeChat, Slack and many more.

11.1.2 Voice Assistants' Introduction

If we consider any technology that can make our life easier in one name, then it is impossible and voice assistant is one of them. A voice assistant is a digital assistant who basically uses recognition of voice, algorithms for language processing and synthesis of voice to listen to the commands of user in audio and revert back to the user with the relevant answer or we can say that to perform the customized task asked by the user. On the basis of some definite commands which is also called intent, that is the voice command from the user, voice assistants can reply with pertinent knowledge by only listening to the specific set of keywords and also by filtering out the surrounding noise [4]. Voice assistants can be generated in the manner that they are software based and which can be further integrated with other hardware technologies as well and others can also use it in their businesses for the growth of the business. But there are some voice assistants that are specifically customized for one hardware and cannot be integrated with any other hardware like Amazon Alexa Wall Clock [5].

Today, voice-assistant technology is unified with many of the devices which we are using on daily basis, for example, mobile phones, computers, smart speakers, cars and much more. Because of their extensive array of amalgamation with other devices, there are multiple voice assistants who provides a very customized set of features; on the other hand, some opt for open-ended situation which can help almost every situation possible.

Technology that is used by voice assistants is Artificial Intelligence (AI) and voice recognition to basically deliver the output for which the user is seeking in a precise and efficient manner. Voice recognition basically works in the manner by taking the input from the user in the form of voice and then transforming it into the analog form which is then turned into the digital signal. After the conversion into the digital form, the computer starts matching the digital signal to the words and the phrases present to recognize the user's intent. But to perform, this computer needs a pre-defined database that consists of these words, phrases, and syllables so that matching can be performed with this database. The technique of matching the digital signal with the existing words and phrases is termed as pattern matching and is the prime force behind voice recognition.

11.1.3 Artificial Intelligence

AI is the key with the help of which human intelligence can be repli-cated by simulating it into machines [6]. In 1950, Alan Turing published "Computing Machinery and Intelligence," which was the first paper or article which questioned that can machines really think? Then after that the Turing test was developed by Alan Turing that checked the capacity of a machine to think like a human and respond accordingly [7]. Later on, four approaches were developed that elaborated AI, thinking humanly/rationally, and acting rationally/humanly [8]. Out of which the initial two deal with reasoning and latter two deal with the actual behavior. Modern AI system is stereotypically seen as a machine which is designed to achieve tasks that generally needs the help of the human interaction [9]. Further, the concept of machine learning (ML) came into picture that improved the simulation of human intelligence into the machines.

11.1.4 Machine Learning

Machine learning denotes to the subset of the AI wherein the programs are generated without the help of the human coders. As a substitute of writ-ing the whole program on their own, programmers feed the AI machine with pattern so that it can be recognized and then can be learnt from it after that, and it gives enormous amount of data to scrutinize through it as well as learn from it. So, in place of having explicit guidelines to stand by, the AI finds the pattern from the database available and the practices it to progress its previously existing functions [10]. One of the ways to make a machine learn and make helpful for voice assistants is by feeding the algorithms hours of speeches that should be in multiple intonations and dialects.

However, traditional functions and programs need input and a set of instructions to process it and produce output. On the other hand, in the case of machine learning programs, we provide them the input and the output and after working on it, they generate the program on its own which maps the input to the output.

Generally, machine learning works on the basis of three approaches: supervised learning, unsupervised learning and reinforcement learning. [11].

11.1.4.1 Supervised Learning

Supervised learning is the basic approach of the ML and it works on the simple concept of training the algorithm with the help of the labeled data [12]. But we have to make sure of one thing that the data which is used in this algorithm should be labeled correctly so that the algorithm can learn on its own but in the correct manner, and this algorithm is super-efficient if it is used in the best circumstances and correct training data.

Supervised learning in ML works by providing it with a small training data set. This data set is the subset of the larger data set and it also assists the algorithm to find out the basic idea of the problem, its probable solution and data points as well that can help the algorithm to deal with it. So basically, the algorithm is given the input in the form of training data which is been already labeled and according to this information, machine makes an algorithm that will satisfy the mapping of the input to the output for the whole training data set and then it will perform on the testing data set and will perform the mapping correctly [13]. So, the solution is then arranged for the usage with the final data set which is been learned from the training data set. And it continues to progress even after determining new patterns and their relationships as it trains on the new data also. So, in supervised learning, we will have a teacher (virtually present) who will tell the algorithm that if the output provided for the given input is correct or not as the data is labeled and if it is not correct, the error will be calculated and it will help the algorithm to improve its performance and will help perform in better manner for the testing data.

11.1.4.2 Unsupervised Learning

Unsupervised learning is the approach in which the ML has the benefit of over the supervised approach of working with the unlabeled data, i.e., there is no need of the extra labor to make the data set machine readable and also allows the larger data set to be worked upon [14]. In contrast to supervised learning approach, unsupervised learning does not contain the labeled data which does not help the algorithm to map the inputs to the outputs, so the hidden structures are created over here that helps in managing the input to the output without the help of labeling of data and this quality of unsupervised algorithm makes it more versatile in nature [15]. So basically, there is no teacher present in this algorithm to tell whether the data mapping is done correctly or not, so a machine learns on its own from the training data that where it should be classified [16].

11.1.4.3 Reinforcement Learning

Reinforcement learning is an approach of learning that takes help from the behavior of humans that how do they learn from the data in their lives [17]. It is an algorithm that improves on its own by learning from new situations and using trial and error method, that is, in this approach, the mostly favorable outputs are reinvigorated and the least favorable ones are neglected or we can say punished [18]. In this algorithm, there is a reward system that is being implemented as there is a teacher present in this approach but it only tells that the output mapped is correct or not. In this method, the error cannot be calculated as the data is gained not labeled, but the reward system helps the approach to identify the favorable outputs and it learns on its own by using the concept of unsupervised learning [19]. In the case of unfavorable outcome of the program, the algorithm is forced to reperform until unless it finds the favorable outcome. And the reward system is directly knotted to the efficacy of the output.

11.2 LITERATURE SURVEY

In [20], it is explained that 2016 was the year of chatbot and it even explained that few commentators mentioned that the chatbots are so unsettling in nature that it will eliminate the need of the websites and applications. As well as the importance of chatbot where it might be useful, what all new technologies are been added. Behavioral health interventions are the challenge for people today and as the market is using voice assistants on daily basis in their needs may it be amazon Alexa, Google Assistant, and Apple's Siri for multiple tasks. It is explained in [21] how the voice assistant technology can be used in behavioral health interventions and Population, Intervention, Comparison, and Outcome (PICO) question are been addressed, then the analysis was carried out. The VA technology is continuously evolving and supporting behavioral interventions which has used multiple platforms like Interactive Voice Response (IVR) systems, smartphones, and smart speakers alone as well as in combination with the others. In [22], research explores that how we can use chatbots in our daily lives, and because of its novelty, it can stand out from the other technologies and it basically explains anticipations for chatbots, preferred input modalities and opportunities for chatbots on the basis of user needs. In [23], talks about how we can inculcate the digital assistance in digital health and digital assistance plays an important role as it is the user interface between the database and the patient. And it specifically explains the landscape of the current integration of the digital health technology in cancer care. In [24], it is explained how Link Student Assistant (LiSA) – a chatbot works, it is organization specific and it helps the student in the campus to retrieve all types of information and services. A behavioral study is also inculcated with the bot properties which then analyses the behavior of the user by the searches. Today, customers are digital friendly and they are involving technology in their day-to-day activities [25] analyzed data from 238 young customers using PLS-SEM which focuses on user's motivation to espouse the digital assistants roleplay in services and this study also helps in providing the managerial aspect for involving intelligent assistants in our day-to-day lives. In [26], review of the chatbot technology is carried out from the United States and India which explains what customer expect from this technology, basic objectives are as following: understand user expectation and perception, surface preferences for input modality and identification of domains where chatbots can be treated as meaningful. In [27], it has been explained that why chatbots are still in evolving zone and not prepared to meet the customers' expectations and how is it helping digital media to feed the consumers as well the clients. In [28], it is explained how chatbots can be efficient in engaging more audience for the museums and galleries businesses through visualizing narrative feature of the chatbot with inclusion of gamification in it. Chatbots are truly expanding business for all sorts of fields. In [29], analysis of two approaches is performed, i.e., pattern matching and ML, and a general architectural design is created for the historical evolution from the day it started until the present time. So, it can act as a very good opportunity for the clients of the history evolution.

11.3 EVOLUTION OF VOICE ASSISTANTS AND ITS TYPES

Digital voice assistants generally work on two approaches first is task-oriented approaches and knowledge-oriented approaches. In order to perform almost every task that a user can put at the voice-activated agent, many voice assistants combine both a task-oriented approach and a knowledge-oriented approach. A **task-oriented approach** is used in a specific goal based on the user's need. Like, make a to-do list, set an alarm and appointment setting, etc., it helps to automate the daily routines of personal. This approach does not require any type of online database to perform the voice command. The task-oriented approach combines itself with other applications to perform a task. While a **knowledge-oriented approach** involves addressing the user's questions like who is the president of a country or today's weather information, a knowledge-based voice assistant accepts the user input in the form of verbal command and processes it with the help of an online database available and helps perform the task. An example of this method is whenever a user asks any question to search on the internet, it will access the online database and find the most relevant results and also provide the highest searched result.

Voice recognition and voice assistant technology are not new. It was invented long before apple Siri in 1960 by IBM. The first voice assistant was developed by IBM with a shoebox device; it was the primary digital voice recognition device that can understand 16 spoken words as well as 0–9 digits. The shoebox voice command-based calculator was presented to the public during the Seattle world's fair in 1962.

A Harpy program was developed at Carnegie Mellon University in Pittsburgh, Pennsylvania with substantial support from the United States Department of Defense and its DARPA agency in the 1970s. This program was capable of understanding 1011 words about the vocabulary just like a three-year-old baby.

Then in the 1990s the very first consumer speech recognition tool was introduced, which was called Dragon Dictate. It was launched at a very high price. And it also launched an updated version called Dragon NaturallySpeaking a few years later. It could understand around 100 continuous spoken words per minute and convert it into the text.

Microsoft launched Clippy (which recalls the paper clip) in 1996, which taught us how to track and interpret natural language in the text to provide us with feedback and suggestions. Being one of the earlier examples of a voice assistant, Clippy enabled speech inputs through Microsoft's Speech Recognition Engine and could be used to help find answers to issues.

In 2010, Siri was introduced by SRI international company with the help of speech recognition provided by Nuance Communications. Initially, it was released as a mobile application on the iOS App Store and acquired by Apple a few months later. After that Siri was released as an integrated part of iOS and first release with iPhone 4s.

After Siri, IBM has announced a new voice assistant named Watson in 2011. Watson was named after the founder of IBM and was originally conceived in 2006 to beat humans at a game of Jeopardy. Watson was one of the most intelligent natural voice recognition systems available at that time.

In 2012, Google launches Google Now features in Google searches. Google added this as a microphone feature to search by voice in google. com. This was the revolutionary step in the voice recognition-based industry. With the help of this functionality, now users can interact with a website through voice and command a website to perform some operation.

After that Microsoft enters the voice assistant-based industry and launched a voice recognition tool Cortana at the annual Build developer conference in 2014. Then, Microsoft includes it in Windows, Xbox, and other Microsoft products as well as in other brands too. Cortana was developed to makeover the operating system for PC and mobile phones. Now Microsoft has to include Cortana in the latest operating system Windows 10 and mobile devices in 2015.

Then, an e-commerce company Amazon launched a smart speaker Amazon Echo with their own digital voice assistant named for the Library of Alexandria in 2014. Alexa was released to assist clients to immerse themselves in the world of home automation. Alexa is probably the most popular voice assistant around and used as a generic term for any voice assistant. Amazon is proactive including voice assistants in a series of smart speakers, mobile phones and other wearables.

In May 2016, Google again came with the spiritual successor of Google Now, and Google Assistant was released with the addition of bi-direction conversation as compared to Google now. Google now returns responses in the form of a Google search result page, while Google Assistant provides responses in the form of natural phrases and with suggestions. Google Assistant is actually a true competitor of Amazon's Voice Assistant Alexa. Google launched a smart voice-activated speaker Google Home with Google Assistant. With the help of a larger tech ecosystem, Google Assistant has become very popular in few days after its debut. The main focus of Google is on making Google Assistant an omnipresent part of the lives of people in daily uses.

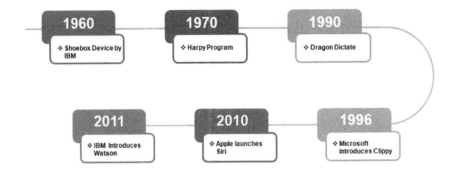

FIGURE 11.2   Evolution of voice assistant from 1960 to 2011.

In 2017, Alan was developed by long-time voice technology's veterans Ramu Sunkara and Andrey Ryabov in Sunnyvale, California. Alan is a highly versatile framework for voice AI built to work with any pre-existing software. You can use the voice assistant to improve the efficiency of any process and because Alan is a fully browser-based integrated development environment, you can edit your scripts on the go whenever the need arises.

In 2017, Samsung debuted in the digital voice assistant arena with Bixby. Bixby 2.0 was developed due to some problems in the initial Bixby. Bixby 2.0 provides a more versatile voice assistant for developers and more customized for consumers than it originally was. Samsung intelligence assistant was first introduced on the Galaxy S8 and S8 and now available in all higher versions.

Companies are working to develop more advanced and more sophisticated voice assistant technology that can help automate more activities and workflows in daily routines, even existing voice assistants such as Siri, Alexa, and Google Assistant learning things with the help of AI. Figure 11.2 explains evolution of voice assistant from 1960 to 2011 and Figure 11.3 explains evolution of voice assistant from 2012 to 2017.

11.3.1 Working of Voice Assistant

With the emergence of virtual assistants such as Alexa from Amazon, Siri from Apple, and Assistant from Google, we get the facility to interact with electronic devices in natural spoken language. Voice assistants are basically a computer program that helps us to talk and provide us with the most relevant outcomes in the form of speech. To begin the conversation with the voice assistant, we need to speak a specific activator keyword like

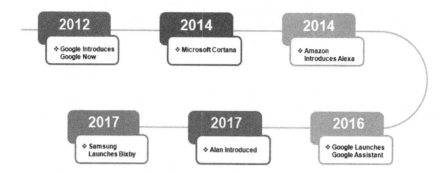

FIGURE 11.3   Evolution of voice assistant from 2012 to 2017.

"Alexa," "Ok Google" and "Hey Siri," etc. After this activation the voice assistant will record our voice with the help of a microphone and process accordingly. There are a number of activities that are needed to resolve a query in voice assistant.

1. On the voice assistant, a wake-word detector or hot word detector runs continuously to listen to a specific activator keyword or phrase to activate the assistant. On mobile devices, we can activate it with a button also.

2. After activation automatic speech recognizer (ASR) converts the input received from the user into text.

3. Furthermore, voice assistants take this text as input and interpret or understand the intention of the user by NLP. A user can make the same request in a different number of ways.

4. A dialogue manager (DM) will be responsible for the response back to the user either it may be any action or conversation. NLP is again used in this part to generate a response in natural language.

5. Text to speech (TTS) is responsible for the response of the query and this response will be delivered to the user in the form of voice through the speaker.

Generally, voice assistants are connected to the internet and online database. The voice assistants use AI and neural network to interpret the query and generate the response in an efficient manner. The voice assistant also

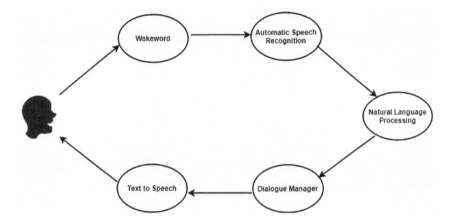

FIGURE 11.4 Working of voice assistant.

uses ML to learn the behavior of the users. So that the voice assistant could understand the future requests made by the user in a better way and perform via speaker. Figure 11.4 explains the working of voice assistant.

11.4 EVOLUTION OF CHATBOTS AND ITS TYPES

Chatbots are the computer program which is known as bots. These bots can communicate with humans through text, voice, and both. In the last few years, chatbots have become very popular due to Siri, Alexa and Google Now are in existence. These chatbots can help automate the daily life routine of a human being efficiently. Conversation based chatbots are not new it was first invented in 1950 by Alan Turing. Alan Turing was an English computer scientist who published a theory of AI and asked "can a machine think." He invented a way to test the ability of chatbots and their intelligence which is called the Turing Test. This was the revolution in the field of chatbots or intelligent assistants that can behave like a human.

In 1966, the very first chatbot "Eliza" was introduced by Joseph Weizenbaum at the MIT laboratory. Eliza was developed to simulate question–answer based conversation with the help of pre-programmed scripted responses. Eliza takes keywords or phrases as input and tries to generate the response by matching the scripted responses.

In 1972, PARRY was developed by Kenneth Colby at the Stanford University. It can behave like a paranoid human and tested against psychiatrists. Only 48% of psychiatrists could detect the difference between PARRY and Human beings.

In 1988, Rollo Carpenter created a chatbot called JABBERWACKY. Basically, it was developed to simulate human conversations in an amusing way. The previous two chatbots were designed for text-only but JABBERWACKY can work on voice also. Rollo Carpenter wanted that it could pass the Turing test but unfortunately, it couldn't.

Dr. Sbaitso was specially developed for MS-DOS in 1992 by Creative Labs. It was a fully voice-based chatbot to communicate with humans. It could behave like a psychologist and chat with a patient in the form of a digital voice. This was the first chatbot that enables AI. Much of his replies were like "Why are you feeling that way?," instead of some form of complex contact.

In 1995, Artificial Linguistic Internet Computer Entity (A.L.I.C.E.) was developed based on heuristic pattern matching to have communicated with humans in human sound. It was more sophisticated than all previous chatbots. It was an application that can run on a computer and formally known as Alicebot. Alicebot was working on Artificial Intelligence Markup Language (AIML). A.L.I.C.E. could not pass the Turing Test but still it was the most advanced chatbot of that time.

In 2001, a chatbot named Smarter Child was developed especially for AOL, MSN, and other messenger networks. It offered many personalized conversations such as information of stock, a score of any sport. The smart child was the precursor to all the commonly accessible virtual assistants we know and love today.

IBM Watson was specially developed for the TV show "Jeopardy!" in 2006. WATSON also ended up defeating a number of previous champions of the show. It showed the strength and potential possessed by conversational agents.

Personal virtual assistants such as Siri in 2010 and Google Now in 2012 was introduced for smartphones. These chatbots could perform tasks on behalf of their human users. Personal assistant chatbots could do anything from placing calls to setting alarm, even browsing the internet, and generate relevant responses.

In 2014, Amazon launched Alexa and Microsoft launched Cortana. Both virtual assistants were capable of voice interaction with human users. Alexa is now built into Amazon Echo smart speaker developed by Amazon. It can perform any web-based task a user can command for. Microsoft Cortana was developed for windows mobile phones and Windows 10 PC. It can also perform web-based activities on behalf of the users. Cortana can help in windows troubleshooting also.

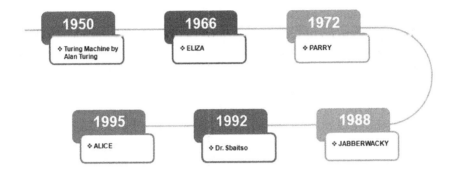

FIGURE 11.5   Evolution of chatbots from 1950 to 1995.

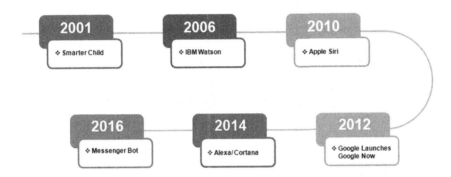

FIGURE 11.6   Evolution of chatbots from 2001 to 2016.

Messenger bot was introduced by Facebook in 2016. It helps users to the Facebook user to be live throughout the day. Messenger bots have improved day by day and provide users with more insights. Figure 11.5 explains the evolution of chatbots from 1950 to 1995, and Figure 11.6 explains the evolution from 2001 to 2016.

11.4.1 Types of Chatbots

There are basically three main types of chatbots on the basis of their learning:

11.4.1.1 Rule-Based Chatbots

This one is the easiest and simplest form of chatbot used today; in these types of bots, people interact with them either by clicking on the buttons or using predefined options. To deliver the best possible output, these chatbots ask the user to make some choices, so there is a possibility that due to limited options it might take time to take the user to the desired goal as they provide options to the user which can be a lengthy process. When anybody aims for the qualification of the leads then it is the best option. So basically, the process works like—A chatbot queries the user and then the user reverts the query with the help of buttons. Then, analysis is done by the bot of the collected data and reverts back to the query. But when it comes to the cutting-edge scenarios which might explore multiple options and situations, this might not be the optimal solution.

11.4.1.2 Intellectually Independent Chatbots
These bots use the concept of ML with the help of which these bots learn through the user's inputs and requests. ML is the concept of the computer which enables to learn on its own from the data with the help of algorithms and then recognizes the pattern to represent the knowledge from the data and takes the least possible interference from the user. The training of intellectually independent chatbots is carried out in the manner to recognize the set of specific keywords and strings that will trigger the bot to request the user query. As time passes by, the bot learns with the integrated concept of unsupervised learning and reinforcement learning that helps it to understand questions and learn their answers for future and always keep learning with the user feedback. So basically, it can be viewed as learning from the experiences and their feedback.

11.4.1.3 AI-Powered Chatbots
AI-powered bots takes the best out of both the above-mentioned techniques and it is an amalgamation of Rule Based Bots and Intellectually Independent Bots. AI is a recreation of human intelligence. It is the cutting-edge technology of computer science that focuses on creating intelligent machines that think and work like humans. AI-powered bots comprehend free language but also make sure to solve the user's problem by following a predefined flow. These bots can memorize the context of the conversation as well as the user choices within the conversation. These bots are smart enough to change the topic of conversation with the flow as well as can address the user request at any instance of time.

11.4.2 NLP and Its Working in Chatbot
NLP is the capacity of the computer to comprehend and analyze human speech, after that to find an appropriate response and reply

in comprehensible human language which can be easily analyzed by humans. The aim of NLP is to make the communication between a machine and human feel like that it is happening between two humans as if there is no presence of a machine. NLP gave the people that freedom of speech between machines and human that they can freely communicate.

NLP involves two sub functions:

a. **Natural Language Understanding (NLU):** It is the capacity of a chatbot to comprehend the human. It is the basic process of translating text into structured data that is easy for any machine to understand.

b. **Natural Language Generation (NLG):** It translates structured data back to the text for basic human understanding.

If we imagine a user asks a chatbot "What is the temperature in Delhi today?" So, to reply back, the bot disrupts the sentence into Entities and Intent. An **Intent** is a request or an action that the user does to get the work done or to get the information he wished for. An **Entity** is the minutiae that praises or intensifies the intent. It can be date, color, size, location, flavor, etc. So, in the above example, **temperature is intent** and the **entities would be Delhi and today.**

There are a few steps which should be followed to complete NLP mechanism, which are as follows:

a. **Tokenization:** In this step, the sentence is broken into tokens that is subparts or words in that sentence.

b. **Part of speech tagging:** Identification of nouns, verbs, adjectives, etc. takes place in this step.

c. **Stemming:** In this step, a word is brought to its basic form by shortening it.

d. **Named entity recognition:** It identifies the entities in the text that user has typed.

e. **Sentiment analysis:** The capacity of the machine to identify the user's emotions as well as the mood.

11.5 UI, DESIGNING, AND TESTING OF INTELLIGENT VIRTUAL ASSISTANTS

For a very long period of time, designers have been creating the GUI, i.e., Graphical User Interface, although volunteering into CUI, i.e., Conversational User Interface, which seems very difficult. CUI is a novel method of implementing human–computer interaction, wherein the mode of communication has changed from graphical elements such as links and buttons to humanly conversations that includes natural language and emotions into consideration. So, CUI is the software that is responsible for performing the conversation between a computer and a human with the help of either text or voice command.

Generally, the steps to keep in mind while designing the software for intelligent agents are as follows.

11.5.1 Acquire Information about the Platform

Intelligent Agent software can be either designed from scratch or from the already existing platform in the form of an app or a website.

On the safer side, a designer can get a chance to design unique UI solutions if the intelligent agent is either brought into the website or the app as there is a freedom to do with the color theme, buttons, links, and design of the existing platform (website or app). But there are other ways with the help of which this software can be created like with the help of the platforms of Telegram, Facebook Messenger, Kik, Slack and many more. They have many restrictions when it comes to visual components. But they do offer perks such as elements, ordering, payments and others. So before going with the designing, the designer should have complete information about the platform.

11.5.2 Enclosing of IA's Purpose

Before starting on the design of the IA, the purpose of the bot should be very clear on how it is going to serve the end user. It's the portion wherein the major roleplay of UX designer comes into picture as it adds the stronger point of view from the user's perspective. Initially after the research on the purpose a set of specifications is fixed but it will evolve with the time span as the user feedback will play a major role in the improvement. Also, the needs and the conversation with the users change with the evolving time.

11.5.3 Establish a Tenor of Voice

Obviously, bots and IAs are not human but still the user prefers the conversations with them to be warm and humanly, that is full of emotions even after knowing that they don't have feelings as it's their main purpose.

To make a bot perform well, it should be researched that what is the bot's end user personality as this personality card will play a major role in establishing the tenor of voice for any bot. If the personality traits have been studied by the designer then only he/she will be able to feed its end user a personality that will satisfy the users need. So, a list should be prepared of keywords and phrases to be used according to the personality traits.

11.5.4 Generate User Flows

Bots and IAs are quite different when it comes to conversation medium as compared to apps and websites because today the majority of the population prefer chatting over mails. So, generating the user flow will help identify the additional features of the bot which can help the end user as well as can provide extra benefits other than texting—maybe a game or something like that for engagement of the user.

11.5.5 Guidance of the Users with UI Design of Chatbot

When it comes to a conversation then a user might ask anything to the bot, so there are endless chances wherein if we fix a set of phrases to initiate a conversation it might not work. So, rather than fixing a set of phrases the designer must check the flow of conversation and should lead the user to the intended conversation for which it is framed. For this procedure, a heuristic approach can be used wherein all possible information should be collected from the user in depth and what is their thought process? what do they want? What should be expected? and other information like this and then it should be fed to the designer so that he can design an appropriate framework. The welcome message can be of great help to make the intentions clear that why is this IA or Bot designed.

Approach should avoid acceptance of open-ended questions as it might lead to the unanswerable mode of bot as well as the closed intent method should be preferred. In addition to this, dead ends can be avoided by giving the user a set of buttons that can lead to a new thread of conversation.

11.5.6 Design for Misinterpretations

The biggest loophole in intelligent agents is the problem of misinterpretation, as few of the bots use the concept of advance NLP and others are still using decision tree logics because no such technology is fixed for it. Because of this, it's not necessary that a bot will always reply in the expected manner and it can lead the conversation to a different direction or maybe a dead end, again the reason being that bots are not completely capable of understanding human emotions which happen even after using the best script.

Having a creative mind to design a bot and also planning of contingency scenarios is the only key solution to the problem of misinterpretations. And the fail conversations should be managed in the manner that it can start a new conversation or can make it meaningful. At last if nothing works out, bots can provide the set of buttons which can initiate the conversation maybe from the start or in different directions.

11.5.7 Trail and Scrutinize User Conduct

The designers always love to get the data from the user for the improvement of the bots, so that the bot's performance can be traced and loopholes can be tracked. There are multiple ways to track a bot's performance like on Botanalytics which will give information of how the bot was used, how it interacted with the user and where did it fail. It can also provide information such as the number of users, the flow of words used mostly and its retention as well as where it failed to answer the user. And this information can also be gained through surveys taken from the users. Q As in questions and buttons can be created to check whether it was satisfactory or not or did it serve the purpose or not. This will be a learning stone for improvisation of the bots for better performance.

11.5.8 Intelligent Agent Testing

There are multiple buzzing words in the market such as intelligent agents, chatbots, voice assistants, virtual agents and many more which interact with humans with the help of machines. Humans communicate with the machines in the form of text or voice commands which is creating a new world of prospects for services such as customer experience, customer support and customer services. The challenge generally faced in intelligent assistants testing is that human interacts in a very assorted manner and so becomes the testing of bot's conversation. Companies generally ensures that they deliver a product of intelligent agent and it should provide the correct information in the minimum time possible and also maintain the human-like nature. The natural language is more diverse and it is really a problem to implement that in the bot and then train accordingly, as well as verification of the data which will lead to the best output. Customers expect a bot to understand in every possible situation, maybe it is their fault that the dialect is not of standard or they might have cold or the voice is not clear, then this becomes the great problem.

After the testing phase, it reveals the weakness, faults also the ability of NLP for not understanding the phrase. In addition it also recommends designing of the chatbot conversations and also how in future both the personality and capability of the bot can be optimized.

11.6 FUTURE SCOPE

When Apple launched Siri in 2011 no one could have expected that this innovation would become a catalyst for software innovation. Almost nine years later we are using virtual assistants either with smart speakers (Google Home with Google Assistant, Amazon Echo with Alexa) or Smartphones to automate our daily tasks. In the near future brands such as Amazon, Google competing for market dominance. Apart from Google and Amazon, many other industries are integrating voice-based virtual assistants in their products to remain in demand.

In the future, virtual assistants will offer streamlined conversations; the virtual assistant will no longer need to use the wake word like "Alexa" or "Ok Google" to activate.

Manufacturers are integrating voice assistant technology into home appliances to make them capable of specific functions. The future VAs will be capable of predicting the best scenario for analytical problems and smooth functioning in workplaces. Intelligent virtual assistants are advancing all kinds of industries ranging from support-based companies to banking, as industries are rushing to release their own voice technology integrations to provide the best service to customer demand.

11.7 SUMMARY

This chapter summarizes about the intelligent virtual assistants that are mainly of two types: voice assistants and chatbots, and their motive is to provide information from the web after the query is provided either in the form of text or in the form of voice. Section 11.1 shares the basic knowledge of the intelligent virtual assistants, i.e., chatbot and voice assistant and the basic working behind them. It also explains the technologies involved in it such as: AI, ML, and types of ML that includes supervised learning, unsupervised learning and reinforcement learning and how these are implemented in different forms of the virtual assistants. Section 11.2 explains the literature survey that is involved in intelligent virtual assistants and how much work has been done in it as well as what are future techniques that will amalgamate with the existing technologies. Then, Section 11.3 talks about the evolution of the voice assistants from its creation to the present, working of the voice assistants as well as the algorithms involved in it and also the types of the voice assistants that are in use today. Section 11.4 explains the evolution of chatbots and their working in depth which involves the briefing of NLP and also the types of chatbots present. Section 11.5 focuses on the UI, Designing and Testing of intelligent virtual assistants in depth and the standards and techniques are involved in it. Section 11.6 explains the future scope of the intelligent virtual assistants and how it is going to impact the society and the whole world.

REFERENCES

1. Bii, Patrick. "Chatbot technology: A possible means of unlocking student potential to learn how to learn." *Educational Research* 4, no. 2 (2013): 218–221.
2. McTear, Michael, Zoraida Callejas, and David Griol. "Creating a conversational interface using chatbot technology." In *The Conversational Interface*, pp. 125–159. Springer, Cham, 2016.
3. Bradeško, Luka, and Dunja Mladenić. "A survey of chatbot systems through a Loebner prize competition." In *Proceedings of Slovenian Language Technologies Society Eighth Conference of Language Technologies*, pp. 34–37. 2012. Ljubljana, Slovenia: Institut Jožef Stefan.
4. Nasirian, Farzaneh, Mohsen Ahmadian, and One-Ki Daniel Lee. "AI-based voice assistant systems: Evaluating from the interaction and trust perspectives." (2017).
5. Bellegarda, Jerome R. "Large-scale personal assistant technology deployment: The siri experience." In *INTERSPEECH*, pp. 2029–2033. 2013.
6. Plant, Robert. "An introduction to artificial intelligence." In *32nd Aerospace Sciences Meeting and Exhibit*, p. 294. 2011.
7. Charniak, Eugene. *Introduction to Artificial Intelligence.* Pearson Education India, 1985.
8. Wenger, Etienne. Artificial intelligence and tutoring systems: Computational and cognitive approaches to the communication of knowledge. Morgan Kaufmann, 2014.

9. Glover, Fred, and Harvey J. Greenberg. "New approaches for heuristic search: A bilateral linkage with artificial intelligence." *European Journal of Operational Research* 39, no. 2 (1989): 119–130.

10. Goodfellow, Ian, Y. Bengio, and A. Courville. "Machine learning basics." *Deep Learning* 1 (2016): 98–164.

11. Ayodele, Taiwo Oladipupo. "Types of machine learning algorithms." *New Advances in Machine Learning* 3 (2010): 19–48.

12. Caruana, Rich, and Alexandru Niculescu-Mizil. "An empirical comparison of supervised learning algorithms." In *Proceedings of the 23rd International Conference on Machine Learning*, pp. 161–168. 2006.

13. Zhu, Xiaojin, and Andrew B. Goldberg. "Introduction to semi-supervised learning." *Synthesis Lectures on Artificial Intelligence and Machine Learning* 3, no. 1 (2009): 1–130.

14. Barlow, Horace B. "Unsupervised learning." *Neural Computation* 1, no. 3 (1989): 295–311.

15. Hastie, Trevor, Robert Tibshirani, and Jerome Friedman. "Unsupervised learning." In *The Elements of Statistical Learning*, pp. 485–585. Springer, New York, NY, 2009.

16. Dayan, Peter, Maneesh Sahani, and Grégoire Deback. "Unsupervised learning." The MIT encyclopedia of the cognitive sciences (1999): 857–859.

17. Sutton, Richard S., and Andrew G. Barto. *Reinforcement Learning: An Introduction*. MIT Press, Cambridge, MA, 2018.

18. Wiering, Marco, and Martijn Van Otterlo. *Reinforcement Learning*. Vol. 12. Springer, Cham, 2012.

19. Sutton, Richard S., and Andrew G. Barto. *Introduction to Reinforcement Learning*. Vol. 135.MIT Press, Cambridge, MA, 1998.

20. Dale, Robert. "The return of the chatbots." *Natural Language Engineering* 22, no. 5 (2016): 811–817.

21. Sezgin, Emre, Lisa K. Militello, Yungui Huang, and Simon Lin. "A scoping review of patient-facing, behavioral health interventions with voice assistant technology targeting self-management and healthy lifestyle behaviors." *Translational Behavioral Medicine* 10, no. 3 (2020): 606–628.

22. Zamora, Jennifer. "Rise of the chatbots: Finding a place for artificial intelligence in India and US." In *Proceedings of the 22nd International Conference on Intelligent User Interfaces Companion*, pp. 109–112. 2017.

23. Garg, Shivank, Noelle L. Williams, Andrew Ip, and Adam P. Dicker. "Clinical integration of digital solutions in health care: An overview of the current landscape of digital technologies in cancer care." *JCO Clinical Cancer Informatics* 2 (2018): 1–9.

24. Dibitonto, Massimiliano, Katarzyna Leszczynska, Federica Tazzi, and Carlo M. Medaglia. "Chatbot in a campus environment: Design of LiSA, a virtual assistant to help students in their university life." In *International Conference on Human-Computer Interaction*, pp. 103–116. Springer, Cham, 2018.

25. Fernandes, Teresa, and Elisabete Oliveira. "Understanding consumers' acceptance of automated technologies in service encounters: Drivers of digital voice assistants adoption." *Journal of Business Research* 122 (2020): 180–191.

26. Zamora, Jennifer. "I'm sorry, dave, i'm afraid i can't do that: Chatbot perception and expectations." In *Proceedings of the 5th International Conference on Human Agent Interaction*, pp. 253–260. 2017.

27. Ask, Julie A., Michael Facemire, Andrew Hogan, and H. B. Conversations. "The state of chatbots." Forrester. com report 20 (2016).

28. Boiano, Stefania, Ann Borda, Guiliano Gaia, Stefania Rossi, and Pietro Cuomo. "Chatbots and new audience opportunities for museums and heritage organisations." Electronic Visualisation and the Arts (2018): 164–171.

29. Adamopoulou, Eleni, and Lefteris Moussiades. "Chatbots: History, technology, and applications." *Machine Learning with Applications* 2 (2020): 100006.

CHAPTER 12

Future Communication Networks

Architectures, Protocols, and Mechanisms for the Next-Generation Internet

Dhananjay Kumar, Sangram Ray, and Sharmistha Adhikari

National Institute of Technology Sikkim

CONTENTS

DOI: 10.1201/9781003121466-12

12.1 INTRODUCTION

The Internet holds a huge part in our daily lives. During 1960s–1970s, in the initial years of its inception its motive was limited to interface a couple of the machines. Since then, the Internet is continually dealing with the enormous needs of the user with the conventional TCP/IP stack-based construction for exchanging the information and rendering different services [1]. The contemporary infrastructure is a host-driven paradigm which works worldwide in the view of correspondence that relies upon both the sender and beneficiaries where the information is just a matter of trade among end-devices. The sender is totally accountable for the correspondence and decisions for exchange of information is centered on the sender's discern. The host-centric Internet design is cumbersome and still precisely abides by design decisions that were taken in accordance with the needs of early Internet users. The circulation of data which is evidently apparent for exchanging data over network has changed its volume massively. Almost all the Internet applications are nothing but the circulation of data from one end device to another end device, for instance, instant messaging, email communication, video conferencing, etc. In view of the past conjectures, Internet congestion came across fast development in the foregoing years. As per the reports by Cisco's Visual Networking Index (VNI), the worldwide Internet congestion has expanded multifold in recent years, and the Compound Annual Growth Rate (CAGR) is projected to reach a growth of 35% during 2015–2019 [2]. The greater part of the congestion is identified with content discovery applications such as IPTV, video streaming via Over the Top (OTT) platforms such as YouTube, Netflix, Hulu, and Video on Demand (VoD). The video data traffic is perceived to solely contribute to almost 86% of all data traffic on the Internet in 2020.

The prerequisite for data transfer speed and bandwidth is expanding day by day [3–9] and demand for user-created contents, Internet Protocol Television (IPTV), and superior quality VoD traffic is continually proceeding at a high development rate by its production using receiver-centric methodologies. The Internet congestion for multimedia contents is ever increasing and the current TCP-/IP-based Internet design isn't adequate to fulfill that sort of content request congestion [10]. The huge growth in the present Internet congestion has created new difficulties in the present IP Internet paradigm. Redundant contents are requested repeatedly causing a lot of Internet congestion. It is important to diminish the replication of identical content requests to make the Internet design proficient [11].

The Internet usage has drastically changed from node-to-node-based delivery to content dissemination irrespective of the location of the content. Customers simply want to think about the content's accessibility and how to get it rapidly and just do not care about the actual serving node of the network. Information exchange can never be out of trend, and there is a need to actualize a proficient Internet design to deal with several Internet concerns which hamper effective information exchange over Internet [12,13].

During the early days of the Internet, it was planned as a military set-up with the objective of reusing and sharing of costly computing assets. Due to the expense and shortage of computing resources, the Internet was then accessed by just a small number of users [14]. These days, the circumstances are divergent, the expense of Internet proficient gadgets have descended, and numerous numbers of people and devices are getting access to it and surprisingly have surpassed a huge number over few billions. Today there exist more than one gadget for every person and this number is projected to grow beyond three gadgets per capita in near future. In any case, the quantity of clients isn't the solitary angle that has carried change to the Internet. The volume of data being disseminated on the Internet has soared hugely along with the number of devices. The Internet congestion for Video-on-Demand (VoD) services will significantly be comparable to the data capacity of 4 billion DVDs in a month. Sadly, the development and advancement of the Internet design and its usage have never considered such increase of Internet clients and the voluminous data transfer over the Internet.

The first Internet design has been kept modifying steadily as and when the requirements for different services rendering was encountered by the Internet service providers. The principal Internet paradigm, Advanced Research Project Agency Network (ARPANET), has associated only an explicit kind of computing device or say, a micro personal computer.

Later on, different sorts of communication networks, for example, satellite and portable device networks, wireless connectivity, are linked through ARPANET to enable communication among more devices. To deal with linkage of devices for information exchange, the TCP/IP protocol stack is formulated and deployed. Furthermore, for lessening the intricacy in discovering terminal devices into interconnected networks, the Domain Name System (DNS) set-up is devised [15].

In spite of its rise, the Internet communication paradigm has stayed unaltered and the network activities are yet driven by an IP address recognizable device-to-device content delivery mechanism. The current communication mechanism of the Internet presents numerous limits on its effective services rendering capabilities and performance in the context of quality of services and the user perceived experiences. In addition, the Internet design offers different duplicates of contents in demand and yet these duplicates are not connected together and can spread inconsistencies. To share expensive computing or other crucial network assets, additional overlay components (say, middleware or protocol translators) have been massively deployed on the top of the TCP/IP protocols stack at appropriate layers. The rapid but credible content exchange entails application-explicit strategic set-ups like Content Delivery Networks (CDN), Peer to Peer (P2P) networking mechanisms for strengthening effective content dissemination architectures [16].

Moreover, security is attained by extrinsic or overlay applications and mechanisms which are mostly supplied by third-party vendors. Trust in discovered content before its dissemination is significantly difficult to accomplish and the greater part of the network linkages depends on unreliable network terminals. Besides these, the reality of discovering the data is closely linked with its storage location's logical identity (IP address). Each content packet compulsorily follows source and destination addressing in a careful manner. Upon content request, packet addresses must be properly labeled. Content requests are firmly mapped to an address to locate the content source, while the requester is only interested in the quick access to the content rather than identifying the serving nodes address. The obvious way out to sort out this concern is to supplant the delivery strategy which was where-centric to something which is what-centric [17]. Communication mechanism built around device-to-device correspondence was a way to tackle the concerns of the 1960s Internet needs, but now it is strongly contended that a communication paradigm must be dependent on named content for improving the present day's need of information exchange.

In light of the aforementioned reasons, host-to-host Internet framework is expected to disappear in near future to develop the scope for one-to-numerous or many-to-many sharing and discovery of data objects with a developing solicitation for an improved architectural scheme for the future Internet. The experiments for more effective future Internet paradigm already have started flourishing across the world [18,19]. Presently the new Internet researchers are emphasizing about the changing over of IP address-centric Internet arrangement to content based and consequently Information-Centric Network (ICNs) set-up is projected. ICN-based Internet designs allies to clean-state architecture of the future Internet [20], in which the content is treated as the prime entity of the entire networking ideology contrary to the present day's Internet in which data packets are tended to as indicated by source and destination terminals IP addresses. ICN-based Internet paradigms route the content as indicated by content name across the network instead of IP addresses of the terminal nodes. Contents are uniquely named and these names serve to reach the content source. The soul emphasis of ICN is to devise, develop, and deploy the content delivery-centric Internet architectures.

The ICN focuses to transform the current TCP/IP protocol stack-based Internet practices to a more generic and capable Internet architecture [21]. The essential system components under an ICN paradigm are no longer the IP address recognizable network devices (computers, switches, routers, hubs, servers, data-centers, etc.) [22,23]. The ICN Internet infrastructure is highly dependent on the named contents, a core feature of ICN. ICN favors a recipient-instigated information exchange paradigm, in which end-client requests a content of their choice by advertising this interest via a well specified Interest message. The whole ICN set-up is responsible for disseminating the named-content request by using its name only in the direction of the best suitable content source and the content is delivered through the converse route to the requesting-client from which the request for the content is coined.

The global society and researchers are interested in clean state Internet infrastructure exploration since design and development of the present Internet are perceived as the constraints. They are constantly trying to redesign the Internet platform by considering the significant requisites and targets as well as releasing new paradigms and prototypes for the future Internet. Keeping these viewpoints in mind, ICN is created as a conceptually ideal postulate for the Future Internet paradigm. The truth of the matter is that the Internet is dynamically reused for data conveyance, moderately less use for start to finish correspondence among end devices. ICN's objective is to give present and future prerequisites improved with respect to the present Internet design. ICN features in-network content caching, to encourage multicast possibilities, consequently conveys an effective discovery and quick delivery of required content to the requesting client [24]. Hence, ICN primarily centers itself around the data and its dispersion. There are numerous existing tasks to actualize the possibility of ICN. ICN usage is led the way at TRIAD [25], which formulated a conceptual structure for persuading ICN researches further. Content-Centric Networking (CCN) [15] is subsequently emerged as Named Data Networking (NDN) [26]. Another ICN implementation is Publish-Subscribe Internet Routing Paradigm (PSIRP), which by passage of time evolved to Publish-Subscribe Internet Technology (PURSUIT) research project [27]. Furthermore, the Network of Information (NetInf) is additionally created to back the future Internet project (4WARD) and its present form is realized in terms of its execution as SAIL project [28]. The architectural designs of these ICN projects are diverse regarding their way of implementation but they share a common fundamental objective behind their different designs. A few prominent ICN Internet paradigms developed so far includes CCN, FP-7 PURSUIT evolved from PURSIP and SAIL Network of Information (NetInf). All of these future Internet perspectives are characterized entrusting utmost importance to fulfill the requesting-client's experience of content retrieval by prolifically focusing on efficacy of content distribution and dissemination.

Among all the ICN Internet architectures, CCN has pulled in impressive consideration from the Internet research community across the globe. The principle concern obstructing the deployment of this architecture is that at least one of these competitive architectures should be leveraged to take the lead to begin as the prime deployment of the core ICN capabilities. These core characteristics of ICN reduces down to two very fundamental aspects of systematic organization of ICN terminals: leveraging in-network storage capability for caching the contents and an ideal name resolution and name-based routing functionality to effectively discover and disseminate the content.

This chapter attempts to present a consolidated view of researches going on for developing an effective, real-life deployable Internet designs. This chapter further discusses the various candidate ICN Internet architectures along with their implementation strategies, strengths and drawbacks. Later on, this chapter also revisits the challenges that are faced by researchers while trying to improve the existing designs of ICN or building a new one from scratch. Last but not the least, a comparative assessment of five ICN perspectives are presented and crucial insights are drawn for open research challenges, vital for strengthening the existing architectures.

This chapter is further structured as follows: Section 12.2 discusses the major challenges that are encountered by the future Internet development communities on the core functional aspects of ICN. The subsequent Section 12.3 presents an extract of five important ICN architectures by quickly revising their key components and functioning. Further in Section 12.4, a brief comparative assessment of important ICN architectures is carried out in terms of their individual implementation strategies, advantages and drawbacks. Finally, Section 12.5 brings forth the conclusive remarks of this chapter.

12.2 ICN RESEARCH CHALLENGES

In this section, prevalent concerns pertaining to ICN architectural development are recognized which have either not been sufficiently tended to or handled by the researchers across the globe. Researches are motivated to explain and expand the ICN way to deal with the future Internet among other directions of the future Internet in light of the ideas, pros and cons discussed in the research outcomes mentioned in which significant give and takes of different Internet researches [24,29] are listed. The various properties and design decisions of the different ICN perspectives are taken based on the challenges posed on its design and deployment. In spite of it, a lot of other significant challenges, which are essential to tackle for the ICN idea to actualize, are to be addressed. The number of the named contents has to be obviously huge as compared to the IP addressable devices connected to the Internet that implies ICN name resolution and name centered content routing framework will have to deal with a massive task to carry out the IP-based routing and DNS name resolution in the real-scale Internet. It incorporates a sophisticated mechanism for aggregation of Interest messages to reduce the information routing overheads and to have expandable name resolution scheme which can be deployed. The ICN awareness about the solicitations for the content are obviously bringing about perhaps more awful security circumstance than exists today for preserving privacy of the content. Then again, it probably won't be conceivable to relate a solicitation to a specific individual. The security concerns should be examined in more detail to comprehend the full outcome and discover intends to moderate them [30]. Pervasive storing likely doesn't sound excessively engaging to a number of content proprietors, who dread that the content published by them can be unlawfully accessed. In that case, the blend of specialized systems, well-built powerful laws and guidelines give an adequate arrangement [31]. The motivations for all stakeholders must be obviously conveyed to develop the required arrangement. It very well may be upheld if different ICN perspectives (e.g., naming, routing, content caching) are duly standardized. ICN's deployment is additionally improved if the ongoing researches permit incremental development and deployment of features and services [32]. Based on the studies of different ongoing ICN research projects [24] the major research challenges in development of ICN architectures can be listed as following:

12.2.1 Scalable Naming Schemes

There is no reasonable agreement yet on whether flat or hierarchical naming schemes ought to be utilized for naming the contents on ICN network. Hierarchical names can be comprehensible and are simpler as a whole to a human; however, it is indistinct whether these aggregated names can expand to the Internet's scale without transforming them using a DNS like set up. Then again, flat names are simpler to manage as it doesn't force preparing necessities for longest prefix discovery. It can act naturally confirming and can be effectively taken care of with exceptionally versatile constructions like Distributed Hash Tables (DHTs) though it is indistinct whether DHTs can offer good implementation benefits or not. There has been for all intents and purposes no examination on fusing, forming, erasure and repudiation of data objects from the naming scheme set-up, and just an inceptive work on the ideal granularity of data objects, like an item might relate to a content, to variable-scale data lumps or to whole application-level articles, have been done. In reality, some work contends that inserting signatures keeps an eye on the individual content but may have inordinate overhead, while other work contends that this is doable with equipment level executions. Looking for data has additionally not gotten a lot of consideration in ICN research, particularly, given that most activities depend upon flat naming schemes there must be some way or another to be found by human clients. Content awareness may give the way to effective data discovery, perhaps taking into account, contextual aspects of content, its type, source, data type, language, and so forth. For instance, SAIL proposes an all-encompassing name scheme that incorporates metadata information to the content retrieval mechanism. In ICN, data is the essential element and it is quite possible for the metadata to coincide with the real data in the network, hence permitting the shrewd control of congestion for different purposes, for example, for empowering geographical routing and stream prioritization. The accessibility of such metadata additionally raises huge concern with respect to lack of biases in the network. Prior endeavors to choke specific sorts of traffic (like, P2P) depended on Dots Per Inch (DPI) procedures. In the ICN, the recognizable proof of data traffic types, and other metadata identified with a stream may establish standard network characteristics and thereby divulging crucial data not exclusively to Internet Service Providers (ISP) but also to network assailants.

12.2.2 Scalable Name Resolution

The huge extent of the naming domain space represents a critical adaptability hindrance for name resolution. Name resolution designs formulated around DHT have pulled in the consideration of scientists because of its logarithmic versatility. The routing strategy infringement and swelled name stretches of DHTs have brought about various hierarchical blueprints that attempt to adjust the design of the naming domain space within the range of inter domain network structure. However, the routing productivity of these methodologies is as yet inadequate. Also, ongoing researches on the design of inter domain linkage propose that increment of peering connections among Asynchronous Servers (AS) forms bit by bit lattice like connection graph. Accordingly, utilizing a carefully hierarchical design for the association of the naming domain space doesn't appear to present the real connections. Another recommended perspective is to utilize hashing to blueprint names straightforwardly to IP addresses and depend on IP-based routing to discover the content sources though this requires worldwide acceptance for being qualified as an effective name resolution framework [33]. Consequently, an adaptable and practically implementable approach, ready to communicate the advancing linkage among ASs, is as yet deficient.

12.2.3 Efficient Name Based Routing

While a great deal of exertion has been given to the blueprint of intra-domain routing instruments, the inter domain routing schemes has received very little consideration [34]. Inter domain routing schemes are firmly influenced by business agreements among the stakeholders and are a matter of dynamic experimentation even with regard to the contemporary design. In the ICN researches, the fundamental concern is mounting the recommended functionalities for the real Internet scale. The inter domain routing by the content routers in NDN face genuine expandability concerns, something that applies to COMET, in which data redirect states are additionally installed at content routers. The PURSUIT variant of ICN paradigms utilizes in-content classic bloom filters for directing routes to content sources and the clearest concern is that the bigger multicast path tree leads to some bogus positives like squandered content circulations. Since larger classic bloom filter paths would present a lot higher overhead, the instruments like classic bloom filter switching and variable-scale classic bloom filter have been investigated [35]. However, the genuine concern is to set up inter domain routes, since it is ridiculous to anticipate from a centralized terminal to have an extensive view on the entire network because of both the size of the Internet and the restricted data traded among ASs. It implies that a various leveled disintegration of the inter domain routing concern is essential among the AS's, combined with classic bloom filter switching, to locally manage the topology and to keep the path stretches short. Furthermore, in the ICN architectures where routes to the content sources are amassed during content name resolution [36], like CONVERGENCE and DONA hybridized with SAIL, fundamental concern remains the vast overhead dispensed during both solicitation and information retrieval due to enlargement of routes to the content sources. MobilityFirst and the standalone versions of SAIL and DONA essentially depend on IP route directing, with a sheer possibility of extra name resolution efforts in the hybridized variant of SAIL and MobilityFirst. This implies that they don't present any new concerns, yet they, somewhat, acquire the current concerns of IP-based routing.

12.2.4 Caching Strategy

Most of the schemes for storing the contents on terminals are in-route storage (and replication) and have been generally devised to be deployed at the application level of the Internet layers. It has been stated that the benefits from the broad practice of caching in ICN won't be considerable. In spite of the fact that they raise genuine worries about the content caching strategies, these perceptions are mostly formulated on the studies conducted many years ago. Additional research on Internet traffic patterns might disclose extra insight to the prevalent attributes of data and consequently to the potential advantages arising from far and wide caching of the contents. For example, a new scrutiny has shown that web data ubiquity got reshaped during the previous couple of years, influencing application layer content caching efficacy and performance. Another concern is that when content caching happens inside the network, a few sorts of data traffic will seek to share the storage space. Thus, effective caching storage management becomes a significant endeavor for the in-network content caching, and further ongoing researches, dependent on streamlined traffic paradigms, have demonstrated that astute content caching designs can considerably improve performance. Furthermore, the in-network installations of storing and replication units opens up the chance of optimizing routing algorithms, data redirect mechanisms and in-network caching simultaneously. For example, content routing choices might be influenced by cache storage locations and the caching capacity of the router as well as sign of storage conflicts.

12.2.5 Congestion Control Mechanisms

The content awareness in ICN designs empowers a progression of systems and in-network capabilities that make information transport a more tedious job as compared to the contemporary. In-network content caching, content replication offers the chance for trading transmission capacity with cache storage capacity, along these lines drastically evolving the transport layer of the network. Besides this, content disseminations modes, for example, one to many (multi-cast) and many to one (con-cast), the capacity of the network to apply any-cast, just as the aid for multi path route directing in a few ICN perspectives, offer a vast set of strategies influencing the pattern of stream, error-control & correction capacities and congestions in the network. Despite all, the way that ICN architectures are as yet under dynamic turn of events, muddles research across globe. Ongoing endeavors have begun to research the cooperation of these strategies, which is anyway a long way from being clearly comprehensible.

12.2.6 Quality of Service (QoS)

Most ICN architecture development activities give significant place to the provisions of Quality of Service (QoS) to be featured in the architecture.

In any case, a couple of ICN architectures give insights concerning functional QoS components, while others treat the concern hastily. The broadest attention to the QoS concerns is in the COMET design which characterizes three groups of services used to provide significant attention to device-to-device content dissemination. COMET delineates the content delivery conveyance necessities of the data as exhibited by group of services into the network routes put forward by every AS through a route provisioning measure. Some similar experiments have been performed on making most of the use of centralized network structure management and routing to the content source in PURSUIT project to actualize content routing scheme that are unattainable with distributed content routing.

12.2.7 Security, Privacy, Access Control, and Trust

Security perspective altogether in ICN designs depends on utilizing encryption with identification keys related to the content names [37]. Very few studies exist anyway on how these identification keys will be overseen, i.e., who are answerable for making, conveying, and renouncing those identification keys. The requirement for the key administration instruments is the fate of foremost significance in the event that might be considered in a way that most of the ICN perspectives depend on cryptographic identification keys and reliable network units for verification of the content names and validation of the content integrity. In addition, many of the recommended ICN designs incorporate access-control instruments, but there is almost no investigation on the meaning of access control strategies, the utilization of the access control arrangements to stored content and the verification of end users authenticity [38]. ICN architectures can be exposed to extreme content privacy protection dangers, as clients coins their curiosity specifically in data and the name of the data being mentioned is accessible to all the ICN network units handling the data solicitation. A persuading answer for this danger has not been given at this point. At last, proficient instruments for building trust connections and taking care of security tussles among the different partners are imagined in ICN architectures [19]; yet this actually stays an open concern.

12.2.8 Mobility

In spite of being recognized as the significant inadequacy of the contemporary infrastructure, network upholds for mobility has received extremely restricted consideration in ICN endeavors [39]. Past research endeavors for featuring mobility into Publish-subscribe architecture and on multicast-backed mobility have significantly pitched up in the direction of comprehending of the arising concerns pertaining to mobility in ICN paradigms. This effort has leveraged to incorporate mobility in PURSUIT paradigm when combined with the indigenous characteristics of ICN paradigm to extend support for in-network content caching and multicast. Still, content and its publisher's mobility stay a significantly difficult task to achieve due to slow update of name resolution systems that comes with the ICN paradigm even though the name-based routing and hierarchical DHTs for name handling exits. The use of the paths to the content sources that may get invalid even as they are framed becomes a stumbling block. Considerably more hazardous is the utilization of requests aggregation based on names in routing tables, as it certainly once again introduces a binding of the content to the location-based identity of the receivers. The most encouraging methodology is the late name mapping supported by hybridized version of SAIL and MobilityFirst, which improve on mobility management [40] the board without depriving of the benefits of the flat naming schemes. However, the efficacy of these naming schemes is uncertain in a large extent real life network.

12.3 IMPORTANT ICN ARCHITECTURES

The diverse ICN architectures [34], whose design initiatives target to replace the current TCP/IP Internet, share some common traits which are important from the point of view of the major functionalities of the next-generation Internet. These common features include content naming, name-centric content routing, in-network caching, mobility, and security. In this section, a brief discussion on five prominent ICN approaches, namely Content-Centric Networking (CCN), Named Data Networking (NDN), Data Oriented Networking Architecture (DONA), Network of Information (NetInf), and Publish-Subscribe Internet Routing Paradigm is presented.

12.3.1 Content-Centric Networking (CCN)

CCN is the underlying venture which was executed at Palo Alto Research Center (PARC) formulated on ICN and is best among the potential ICN projects undertaken by Jacobson et al. in 2009 [10]. The purpose of CCN [41] is to substitute the IP address-centric Internet blueprint to the named-content-centric blueprint. The unit of data is chosen as content chunks in CCN, and is named according to a hierarchical naming design which follows common prefix structure as in, ccnx:/parc.org/video/. Interest messages and Data messages are the two native constructs which ensure that content correspondence is conceivable [42]. Interest messages signify the customer's solicitations while data message manifests the reply from content source. Interest message matches with the data message as the name prefix of interest message when a data hit happens. CCN core entities are separated into different data structures which includes: a Content Store (CS), a Pending Interest Table (PIT) and a Forwarding Information Base (FIB) [43]. Communicated contents on the delivery routes are stored at intermediate CSs. PIT deals with all the solicitations and determined routes for serving them with the content. FIB demonstrates the potential routes for Interest message to advance toward the suitable content source and also forward the message containing the content from the content source to the requesting client. The functioning of CCN design is delineated in Figure 12.1. At the point when customers send their solicitations for required content then the solicitations sent to the CCN nodes that are closest to the mentioned client. At that point, the solicitation is contrasted with the current content in CS. On the off chance that the solicitation doesn't matches with any of the current content in CS the CS then forward the solicitation toward the content sources. As soon as the desired content

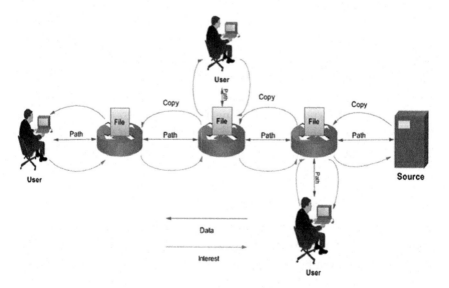

FIGURE 12.1   Content-Centric Networking (CCN).

is discovered, the content is reverted to the requesting client through a cached replica and a duplicate likewise of the content is also stored at the intermediate CCN nodes that show up along the content conveyance route back to the requesters.

12.3.2 Named Data Networking (NDN)

NDN project was undertaken and driven in Los Angeles, at the University of California, along with active cooperation from around 10 other institutions and Internet research establishments from United States. The underlying thought of the undertaking this project is rooted back to the idea of content-centric networks (CCNs) initiated by Ted Nelson during 1970s. Since then, a few endeavors, like, in Berkeley, DONA, at University of California and TRIAD, at Stanford were completed investigating networking based on the named contents. Xerox Palo Alto Research Center (PARC) in 2009 delivered the CCNx project carried on by Van Jacobson. He was a pioneer and one among the other specialized heads for the NDN project. The essential contention of the NDN project is that the contemporary use of the Internet has changed the device-to-device data transmission to the content-driven paradigm. The contemporary Internet, which is a client-server paradigm, is confronting difficulties in tendering secure and managed content deliveries [44]. In this data-centric paradigm, the network is straightforward and simply sending information, i.e., it is content-ignorant. Because of this ignorance, numerous duplicates of similar information are sent between end devices on the network over and over with no traffic improvement on the network's part. The NDN utilizes an alternate paradigm that empowers the network to focus on what content instead of where is the content located. The information is named rather than their bothering about their storage location. Content becomes the top-notch element in NDN. Rather than attempting to get the transmission channel or information in transit secured using encryption, the NDN attempts to secure the content by adopting a security-improved strategy for naming the contents. The methodology permits isolating trust in information from trust among clients and servers, which may significantly empower content reserving on the network routes to upgrade traffic of data. Figure 12.2 presents a straightforward delineation of the objective of the NDN paradigm to assemble a tight midsection around contents rather than IP addresses. NDN widely opens to a few prime exploration concerns. The initial concern is the means by which to discover the

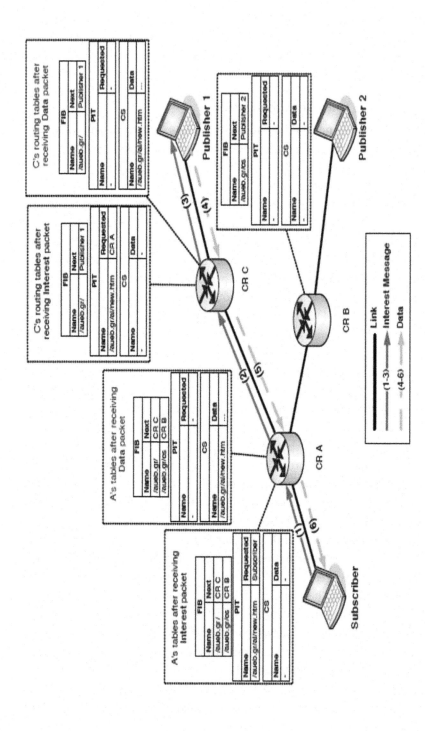

FIGURE 12.2   Named Data Networking (NDN).

information, or how the information is named and coordinated to guarantee quick information query and conveyance. The recommended thought is to adopt a various leveled name-structure name the contents, which is adaptable and simple to recover. The subsequent examination concern is information security and reliability. NDN proposes to get the information secured straightforwardly as opposed to getting the information compartments secured, like records, files, and network linkage. The contents are endorsed by public identification keys. The third concern is the expandability of NDN. NDN names are longer than IP addresses, yet the various leveled name structure may improve the effectiveness of query and worldwide availability of the information. As to concerns, NDN attempts to address them en-route to determine the difficulties in content routing, adaptability, trust and security paradigms, quick information discovery and conveyance, content integrity and protection, and a fundamental hypothesis supporting the blueprint.

12.3.3 Data Oriented Networking Architecture (DONA)

DONA is another clean-state ICN engineering which is formulated on the idea of location independent content delivery [45]. It is the first functionally absolute ICN design that was set up to changes over the host-to-host-based network into the information-centric network. Hierarchical naming scheme is utilized in the DONA set-up and the unit of data is expressed using Named Data Objects (NDOs). Every data entity is exceptionally recognized by its name coupled with a public private identification key pair, where the public identification key represents the distributor identity and private key is a label marked out by the distributor on the content. Route directing of NDOs is accomplished through the help of Resolution Handlers (RHs). Every DONA node consists of an enlisted data indexing table which stores three sorts of data: first is the name of the NDO in public-private key pair second is the next expected RH to think about the contents accessibility and the third one is the distance to next hope. Customer's solicitation and corresponding information from the source is sent utilizing two methodologies: REGISTER and FIND. As soon as the RH gets client's demand, the each solicitation message is contrasted with existing names of enlisted data indexing table, on the off chance that the solicitation doesn't matches with any currently registered names; the solicitation is sent to the relating next intended RH as demonstrated in Figure 12.3. The information is reverted to the requesting through the

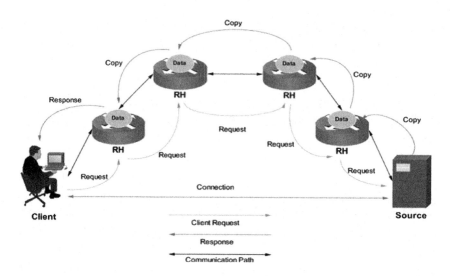

FIGURE 12.3 Data Oriented Networking Architecture (DONA).

opposite route when the necessary content is found and a replica of the content is stored at the return route routers or the data is sent directly through the dedicated route [46] as demonstrated in Figure 12.3. In the event that the enlisted data table contains more than one passage, demand is shipped off the closest wellspring of the necessary content.

12.3.4 Network of Information (NetInf)

The next ICN paradigm to actualize the new age Internet architecture is the NetInf. It is the distinct division of another new age Internet undertaking called 4WARD [47] which gets its roots from the one another similar project widely known as Scalable and Adaptive Internet Solutions (SAIL). NetInf joins the information-level data and bit-level data entities into a single entity called Named Data objects (NDOs), similar to DONA. NDOs are characterized through three constituent units namely, a universally unique content name, the metadata information carrying the semantic and other essential aspects of the content and a hash function built-in for security capacities. The bit-level object is the actual content which contains the data in its binary representation of the actual content. The flat naming scheme is utilized in NetInf as a dual of the public identification key of the content supplier and the content label assigned by the content providers. The NDOs are first enrolled in the Name Resolution System (NRS) using their location identifiers when the publishers distribute their contents.

All customer demands is received and handled by NRS, and an extensive matching is performed by NRS [29] among the names indicated in solicitations and listed in existing NDOs. The NRS then sends the solicitation to the most suitable content source if there is any passage identified by contrasting with the name in the solicitation message. Figure 12.4 clarifies how the customer's solicitation forward across the network as indicated by NRS. In NetInf ICN design, all contents entities are enlisted in NRS and the NRS ceaselessly keeps updating itself from time to time. NRS performs content name matching among the currently existing NDOs and the name indicated in demands for content. In the event that there exists any of the NDO identified with the name in the demand the NRS redirects the solicitation messages toward the most suitable content source. As soon as solicitation is received at source, the source responds back to the requesting client through the converse route or via the dedicated route as demonstrated in Figure 12.4.

12.3.5 Publish-Subscribe Internet Routing Paradigm (PSIRP)

The project Publish-Subscribe Internet Routing Paradigm (PSIRP) is inspired by the well-known Publish-Subscribe communication design. The inception of PSRIP is rooted in Publish-Subscribe Internet Technology (PURSUIT) project. Both these designs are the piece of European FP7 venture which was begun in 2008 and finished in 2012 [35]. Both of these ventures were acquainted to actualize and approve the proposal of

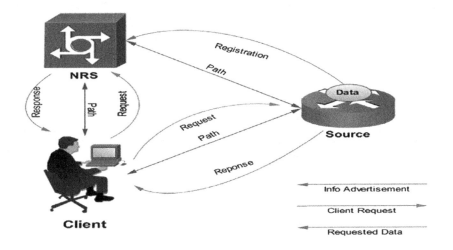

FIGURE 12.4 Network of Information (NetInf).

data-driven networking, i.e., ICN. The PSIRP utilized Rendezvous framework to deal with naming and routing prerequisites to incorporate global uniqueness in content names and to broaden the scope of routing. This Rendezvous framework is separated into a few Rendezvous Nodes (RN) and those nodes are further interlinked to make content correspondence conceivable. Two sorts of identifiers popularly recognized as Rendezvous identifier (Rid) and the Scope identifier (Sid) are utilized to distribute all the contents across the network. Rids and Sids are utilized to characterize the globally unique content name and the worldwide extent of content. The mechanism of Scope identification is brought in to hierarchically structure the information and the identifiers express novel names and the scope for all contents. The RN is answerable for performing coordinating between the distributor's content and the customer demand. Subsequent to content discovery the Forwarding Node (FN) is mindful to advance the content closer to the requesting clients from the content source. The FN comprises the classic bloom filters which are utilized to distinguish a particular node interface by which the mentioned content requirements are to be served. The membership demand aggregates the return route in the classic bloom filter and generates the corresponding Fid, a bloom filter storage which stores links to the route segments. By utilizing the Fid the necessary content is reverted to the requester. Figure 12.5 clarifies the core design of rendezvous in PURSIP. The source rendezvous and

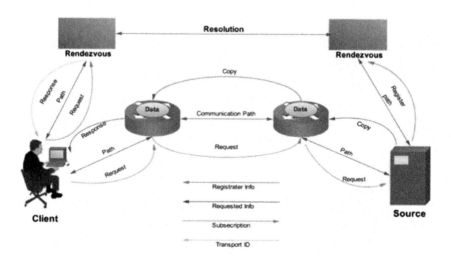

FIGURE 12.5   Public-Subscribe Internet Routing Paradigm (PSIRP).

client rendezvous are interconnected to exchange information with one another. The customer needs to send the solicitation to the corresponding rendezvous. In the event that the solicitation matches with any of the currently available listing, the solicitation is shipped off the source which is referenced in the rendezvous. Followed by which a duplicate of the mentioned information is replicated at the conveyance route from source to the requester as demonstrated in Figure 12.5.

12.4 COMPARATIVE ANALYSIS OF ICN ARCHITECTURES

The aforementioned ICN paradigms share few likenesses and at the same time contrasts in the manner they actualize the key ICN functionalities. In this section of this chapter, we give a brief but fundamental examination and correlation of the distinctive design decisions of these ICN architectures. Table 12.1 summarizes the comparative distinction among the five ICN paradigms discussed earlier.

12.5 CONCLUSION

This chapter attempts to give a top to bottom study of the ICN research initiatives in seek of an effective future Internet architecture. As an initial step, a series of concerns are recognized in the current Internet design. These concerns are taken as key perspectives of how the Internet ought to work for adapting the new and arising necessities. Further it follows with the list of concerns that can be encountered while proceeding with ICN research. A few ICN designs have been briefly discussed to get the reader an idea of the fact that how these ICN perspectives address some of today's Internet necessities, like the requirement for productive data conveyance and portability uphold. It is seen that how ICN research has marked its presence in the most recent decade, with a significant sprout of related exercises occurring during the Internet evolution. Despite the fact that the ICN related researches are yet molding, this chapter put forth an attempt to give a brought together perspective on the elective recommendations by characterizing a bunch of central ICN functionalities, e.g., naming, name resolution, name-based content routing, in-network content caching, portability, and security. Further this chapter introduces five ICN architectures, clarifying their overall objectives and activity, just as how they execute every one of the core ICN functionalities, coming with quick comparative analysis of the different design decisions with respect to the core ICN characteristics.

TABLE 12.1 Comparative Analysis of Different ICN Architectures

ICN Architectures	Implementation Strategies	Advantageous Features	Drawbacks
CCN [10, 35]	Hierarchical naming, human readable names, signature-based security for solicitation messages, name-based routing of interest packet, reverse Interest path routing for data and Interest and Data packets.	Inherent any-cast support, recirculation support via in-network packet caching and content awareness in caching and routing Decisions.	None guaranteed content delivery, Connection breakage, PIT flooding, Clog due to the PIT table size, Colossal intricacy added to the routing aggregation, No assurance of the content retrieval, Not expandable for large-scale inter domain communication.
NDN [24, 35]	Hierarchical naming, name resembles uniform resource locators, human readable names, name-based routing to flood names, signature-based security for solicitation messages, reverse Interest path routing for data and Interest and Data packets.	Quest for data object by exploring hope by hope, dependable, low dormancy and worldwide conveyance, supports full name interest packets and debilitate the interest packet loops with the help of request aggregation.	Mostly similar to CCN, Connection breakage, PIT flooding, Clog due to the small PIT table size, Colossal intricacy added to the routing aggregation, No assurance of the content retrieval, Not expandable for large-scale inter domain communication.

(Continued)

TABLE 12.1 (CONTINUED) Comparative Analysis of Different ICN Architectures

ICN Architectures	Implementation Strategies	Advantageous Features	Drawbacks
DONA [35,41]	Flat naming scheme, human unreadable names, signature-based security, fit on key performance metrics, name-based routing assisted by resolution handlers, reverse Interest path and direct path routing for data and named data objects.	Resolution mechanism is assisted with resolution handlers (RHs), one logical RH for each domain and operational primitives consist of REGISTER, FIND and PUSH message supports.	Large scale expandability challenge, Inter-domain routing concerns, Centralized failure concern with RHs, High routing overheads drew on hope distance between requester and content source, Complex inter-domain linkage graphs.
NetInf [35]	Flat naming, human unreadable names, signature-based security, hash function, fit on Key performance metrics, named data objects, named-based resolution assisted by name resolution system and classic bloom filter-based data routing.	Supports the standalone and hybridized content dissemination, structured along a two level global and local distributed hash tables mechanism and incorporates data exchange via PUBLISH, GET, and DATA messages.	High operational overheads, centralized node failures concerns, additional name resolution efforts and not expandable to large-scale Internet needs due to complex management of a huge scale naming domain space which cannot be aggregated.
PSIRP [35, 37]	Flat naming, human unreadable, signature-based security, name-based routing assisted by Rendezvous, reverse path or direct path routing and named data objects.	Uses in-packet classic bloom filters for content source routing, Supports distributed hash table-based rendezvous network, Routing strategy is made up of 4 parts; Rendezvous, routing, network structure, and data redirect.	Large scale expandability concerns, longer paths resulting in bogus positives, centralized failure concerns, problematic Inter domain routing, huge storage requirements for storing NDOs and frequent packet circulation wastage.

At last, this chapter expresses that ICN is a promising and rich innovation field that has shown its potential for tending to probably a vast portion of the current concerns of the Internet. There is, accordingly, a pressing requirement for additional exploration and quantitative examinations for assessing the advantages and potential execution gains brought by this new Internet design paradigm. Along the way, this chapter has clearly pointed out the open concerns for ICN structures, with respect to its core functionalities to leverage the readers to continue their curiosities in further evolution of ICN architectures.

ACKNOWLEDGMENT

This work is supported by the R&D project grant no. DST/ICPS/CPS-Individual/2018/806(G) funded by the Department of Science and Technology, Government of India.

REFERENCES

1. Alberti, Antonio Marcos, Marco Aurelio Favoreto Casaroli, Dhananjay Singh, and Rodrigo da Rosa Righi. "Naming and name resolution in the future internet: Introducing the NovaGenesis approach." *Future Generation Computer Systems* 67 (2017): 163–179.
2. Forecast, G. M. D. T. "Cisco visual networking index: Global mobile data traffic forecast update, 2017–2022." *Update* 2017 (2019): 2022.
3. Fazea, Yousef. "Numerical simulation of helical structure mode-division multiplexing with nonconcentric ring vortices." *Optics Communications* 437 (2019): 303–311.
4. Fazea, Yousef. "Mode division multiplexing and dense WDM-PON for Fiber-to-the-Home." *Optik* 183 (2019): 994–998.
5. Fazea, Yousef, Mustafa Muwafak Alobaedy, and Zeyid T. Ibraheem. "Performance of a direct-detection spot mode division multiplexing in multimode fiber." *Journal of Optical Communications* 40, no. 2 (2019): 161–166.
6. Fazea, Yousef, and Angela Amphawan. "5× 5 25 Gbit/s WDM-MDM." *Journal of Optical Communications* 36, no. 4 (2015): 327–333.
7. Fazea, Yousef, and Vitaliy Mezhuyev. "Selective mode excitation techniques for mode-division multiplexing: A critical review." *Optical Fiber Technology* 45 (2018): 280–288.
8. Fazea, Yousef, and Angela Amphawan. "32 channel DQPSK DWDM-PON for local area network using dispersion compensation fiber." In *EPJ Web of Conferences*, vol. 162, p. 01016. EDP Sciences, InCAPE2017, 2017.
9. Hassan, Suhaidi, Yousef Fazea, Adib Habbal, and Huda Ibrahim. "Twisted Laguerre-Gaussian mode division multiplexing to support blockchain applications." In *TENCON 2017-2017 IEEE Region 10 Conference*, pp. 2421–2050. IEEE, Penang, Malaysia, 2017.

10. Naeem, Muhammad Ali, Shahrudin Awang Nor, Suhaidi Hassan, and Byung-Seo Kim. "Performances of probabilistic caching strategies in content centric networking." *IEEE Access* 6 (2018): 58807–58825.

11. Naeem, Muhammad Ali, Shahrudin Awang Nor, and Suhaidi Hassan. "Future Internet Architectures." In *International Conference of Reliable Information and Communication Technology*, pp. 520–532. Springer, Cham, 2019.

12. Naeem, Muhammad Ali, Rashid Ali, Byung-Seo Kim, Shahrudin Awang Nor, and Suhaidi Hassan. "A periodic caching strategy solution for the smart city in information-centric Internet of Things." *Sustainability* 10, no. 7 (2018): 2576.

13. CV Networking. CiscoGlobal Cloud Index: Forecast and methodology, 2015–2020. White paper by Cisco (2017).

14. Dräxler, Martin, and Holger Karl. "Efficiency of on-path and off-path caching strategies in information centric networks." In *2012 IEEE International Conference on Green Computing and Communications*, pp. 581–587. IEEE, Nov. 20 2012 to Nov. 23 2012 Besancon, France, 2012.

15. Ahlgren, Bengt, Christian Dannewitz, Claudio Imbrenda, Dirk Kutscher, and Borje Ohlman. "A survey of information-centric networking." *IEEE Communications Magazine* 50, no. 7 (2012): 26–36.

16. Singhal, Rahul, Eric K. Geyer, Henning Makholm, and Christian Worm Mortensen. "Marker based reporting system for hybrid content delivery network and peer to peer network." U.S. Patent 10,887,385, issued January 5, 2021.

17. Lu, Zhihui, Ye Wang, and Yang Richard Yang. "An analysis and comparison of CDN-P2P-hybrid content delivery system and model." *JCM* 7, no. 3 (2012): 232–245.

18. Montpetit, Marie-Jose, Serge Fdida, and Jianping Wang. "Future internet: Architectures and protocols." *IEEE Communications Magazine* 57, no. 7 (2019): 12–12.

19. Gavras, Anastasius, Arto Karila, Serge Fdida, Martin May, and Martin Potts. "Future internet research and experimentation: The FIRE initiative." *ACM SIGCOMM Computer Communication Review* 37, no. 3 (2007): 89–92.

20. Yan, Huan, Deyun Gao, Wei Su, Chuan Heng Foh, Hongke Zhang, and Athanasios V. Vasilakos. "Caching strategy based on hierarchical cluster for named data networking." *IEEE Access* 5 (2017): 8433–8443.

21. Pan, Jianli, Subharthi Paul, and Raj Jain. "A survey of the research on future internet architectures." *IEEE Communications Magazine* 49, no. 7 (2011): 26–36.

22. Fisher, Darleen. "A look behind the future internet architectures efforts." *ACM SIGCOMM Computer Communication Review* 44, no. 3 (2014): 45–49.

23. De Meer, Hermann, Karin Anna Hummel, and Robert Basmadjian. "Future Internet services and architectures: Trends and visions." *Telecommunication Systems* 51, no. 4 (2012): 219–220.

24. Paschos, Georgios S., George Iosifidis, Meixia Tao, Don Towsley, and Giuseppe Caire. "The role of caching in future communication systems and networks." *IEEE Journal on Selected Areas in Communications* 36, no. 6 (2018): 1111–1125.

25. Fang, Chao, Haipeng Yao, Zhuwei Wang, Wenjun Wu, Xiaoning Jin, and F. Richard Yu. "A survey of mobile information-centric networking: Research issues and challenges." *IEEE Communications Surveys & Tutorials* 20, no. 3 (2018): 2353–2371.

26. Arianfar, Somaya, Pekka Nikander, and Jörg Ott. "On content-centric router design and implications." In *Proceedings of the Re-architecting the Internet Workshop*, pp. 1–6. 2010. ReArch'10 was held in conjunction with 6th ACM International Conference - Co-NEXT '10: Conference on emerging Networking Experiments and Technologies, Philadelphia, PA, 30 November 2010.

27. Chai, Wei Koong, Diliang He, Ioannis Psaras, and George Pavlou. "Cache "less for more" in information-centric networks (extended version)." *Computer Communications* 36, no. 7 (2013): 758–770.

28. Cui, Xian-dong, Tao Huang, L. I. U. Jiang, L. I. Li, Jian-ya Chen, and Yun-jie LIU. "Design of in-network caching scheme in CCN based on grey relational analysis." *The Journal of China Universities of Posts and Telecommunications* 21, no. 2 (2014): 1–8.

29. Naeem, Muhammad Ali, Shahrudin Awang Nor, Suhaidi Hassan, and Byung-Seo Kim. "Compound popular content caching strategy in named data networking." *Electronics* 8, no. 7 (2019): 771.

30. Naeem, Muhammad Ali, and Suhaidi Hassan. "IP-Internet data dissemination challenges and future research directions." *Journal of Advanced Research in Dynamical and Control Systems* 10, no. 10 (2018): 1596–1606.

31. Bechtold, Stefan, and Adrian Perrig. "Accountability in future internet architectures." *Communications of the ACM* 57, no. 9 (2014): 21–23.

32. Peng, Chen, Qing-Long Han, and Dong Yue. "Communication-delay-distribution-dependent decentralized control for large-scale systems with IP-based communication networks." *IEEE Transactions on Control Systems Technology* 21, no. 3 (2012): 820–830.

33. Azgin, Aytac, Ravishankar Ravindran, and Guoqiang Wang. "Hash-based overlay routing architecture for information centric networks." In *2016 25th International Conference on Computer Communication and Networks (ICCCN)*, pp. 1–9. IEEE, Waikoloa, HI, 2016.

34. Saha, Sumanta, Andrey Lukyanenko, and Antti Ylä-Jääski. "Cooperative caching through routing control in information-centric networks." In *2013 Proceedings IEEE INFOCOM*, pp. 100–104. IEEE Computer and Communications Societies, Turin, Italy, 2013.

35. Xylomenos, George, Christopher N. Ververidis, Vasilios A. Siris, Nikos Fotiou, Christos Tsilopoulos, Xenofon Vasilakos, Konstantinos V. Katsaros, and George C. Polyzos. "A survey of information-centric networking research." *IEEE Communications Surveys & Tutorials* 16, no. 2 (2013): 1024–1049.

36. Barakabitze, Alcardo Alex, Tan Xiaoheng, and Guo Tan. "A survey on naming, name resolution and data routing in information centric networking (ICN)." *International Journal of Advanced Research in Computer and Communication Engineering* 3, no. 10 (2014): 8322–8330.

37. Fu, Xiaoming, Dirk Kutscher, Satyajayant Misra, and Ruidong Li. "Information-centric networking security." *IEEE Communications Magazine* 56, no. 11 (2018): 60–61.

38. Ambrosin, Moreno, Alberto Compagno, Mauro Conti, Cesar Ghali, and Gene Tsudik. "Security and privacy analysis of national science foundation future internet architectures." *IEEE Communications Surveys & Tutorials* 20, no. 2 (2018): 1418–1442.

39. Adhatarao, Sripriya, Mayutan Arumaithurai, Dirk Kutscher, and Xiaoming Fu. "NeMoI: Network mobility in ICN." In *International Conference on Communication Systems and Networks*, pp. 220–244. Springer, Cham, 2018.

40. Kitagawa, Taku, Shingo Ala, Suyong Eum, and Masayuki Murata. "Mobility-controlled flying routers for information-centric networking." In *2018 15th IEEE Annual Consumer Communications & Networking Conference (CCNC)*, pp. 1–2. IEEE, 12–15 January 2018, Las Vegas, 2018.

41. Adhikari, Sharmistha, Sangram Ray, Gosta P. Biswas, and Mohammad S. Obaidat. "Efficient and secure business model for content centric network using elliptic curve cryptography." *International Journal of Communication Systems* 32, no. 1 (2019): e3839.

42. Adhikari, Sharmistha, and Sangram Ray. "A Lightweight and secure IoT communication framework in content-centric network using elliptic curve cryptography." In *Recent Trends in Communication, Computing, and Electronics*, pp. 207–216. Springer, Singapore, 2019.

43. Adhikari, Sharmistha, Sangram Ray, Mohammad S. Obaidat, and G. P. Biswas. "Efficient and secure content dissemination architecture for content centric network using ECC-based public key infrastructure." *Computer Communications* 157(2020): 187–203.

44. Adhikari, Sharmistha, and Sangram Ray. "A secure anonymous mobile handover authentication protocol for content centric network." In *International Conference on Computational Intelligence, Communications, and Business Analytics*, pp. 360–373. Springer, Singapore, 2018.

45. Vasilakos, Athanasios V., Zhe Li, Gwendal Simon, and Wei You. "Information centric network: Research challenges and opportunities." *Journal of Network and Computer Applications* 52 (2015): 1–10.

46. Kutscher, Dirk, Suyong Eum, Kostas Pentikousis, Ioannis Psaras, Daniel Corujo, Damien Saucez, Thomas C. Schmidt, and Matthias Waehlisch. "Information-centric networking (ICN) research challenges." Internet Research Task Force (IRTF) (2016):1–38.

47. Qazi, Faiza, Osman Khalid, Rao Naveed Bin Rais, and Imran Ali Khan. "Optimal content caching in content-centric networks." Wireless Communications and Mobile Computing 2019 (2019).

Index

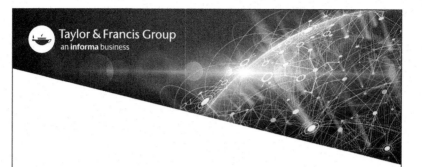

Taylor & Francis eBooks

www.taylorfrancis.com

A single destination for eBooks from Taylor & Francis
with increased functionality and an improved user
experience to meet the needs of our customers.

90,000+ eBooks of award-winning academic content in
Humanities, Social Science, Science, Technology, Engineering,
and Medical written by a global network of editors and authors.

TAYLOR & FRANCIS EBOOKS OFFERS:

A streamlined
experience for
our library
customers

A single point
of discovery
for all of our
eBook content

Improved
search and
discovery of
content at both
book and
chapter level

REQUEST A FREE TRIAL
support@taylorfrancis.com

 Routledge
Taylor & Francis Group

 CRC Press
Taylor & Francis Group

Printed in the United States
by Baker & Taylor Publisher Services